SpringerBriefs in Water Science and Technology

SpringerBriefs in Water Science and Technology present concise summaries of cutting-edge research and practical applications. The series focuses on interdisciplinary research bridging between science, engineering applications and management aspects of water. Featuring compact volumes of 50 to 125 pages (approx. 20,000–70,000 words), the series covers a wide range of content from professional to academic such as:

- Literature reviews
- In-depth case studies
- Bridges between new research results
- Snapshots of hot and/or emerging topics

Topics covered are for example the movement, distribution and quality of freshwater; water resources; the quality and pollution of water and its influence on health; and the water industry including drinking water, wastewater, and desalination services and technologies.

Both solicited and unsolicited manuscripts are considered for publication in this series.

Manish Kumar Goyal · Sachidanand Kumar ·
Akhilesh Gupta

AI Innovation for Water
Policy and Sustainability

 Springer

Manish Kumar Goyal ⓘ
Department of Civil Engineering
Indian Institute of Technology
Indore, Madhya Pradesh, India

Sachidanand Kumar
Department of Civil Engineering
Indian Institute of Technology
Indore, Madhya Pradesh, India

Akhilesh Gupta
Goverment of India
Department of Science and Technology
New Delhi, Delhi, India

ISSN 2194-7244   ISSN 2194-7252   (electronic)
SpringerBriefs in Water Science and Technology
ISBN 978-3-031-72013-0   ISBN 978-3-031-72014-7   (eBook)
https://doi.org/10.1007/978-3-031-72014-7

This Springer imprint is published by the registered company Springer Nature Switzerland AG
The registered company address is: Gewerbestrasse 11, 6330 Cham, Switzerland

If disposing of this product, please recycle the paper.

Preface

In recent years, the intersection of artificial intelligence (AI) innovation and water policy has emerged as a critical area of focus in the quest for sustainable water management practices. As the world's population expands and urbanizes, the need for water resources increases, presenting unprecedented challenges for policymakers and water managers globally. In response to these challenges, the integration of AI technologies offers immense potential to revolutionize how we approach water policy and sustainability.

This book explores into the multifaceted relationship between AI innovation and water policy, exploring the latest advancements, case studies, and emerging trends shaping the field. Chapter 1 introduces the basics of AI for water management. Chapter 2 explains the AI for water conservation. Chapter 3 describes the AI for water treatment. Chapter 4 deals with the AI for water policy. Chapter 5 provides the framework for future AI for future water. Several sources have been referenced to gather relevant and comprehensive information for this book, and the authors acknowledge them.

By providing a platform for dialogue and knowledge exchange, this book aims to inspire policymakers, researchers, and stakeholders to harness the full potential of AI innovation for shaping a more water-secure and sustainable future.

Indore, India   Manish Kumar Goyal
Indore, India   Sachidanand Kumar
New Delhi, India   Akhilesh Gupta

Acknowledgments

Writing this book on AI Innovation for Water Resource Management and Sustainability has been a collaborative effort, and we are deeply grateful to everyone who has contributed to its realization. We would also like to thank the editorial team and staff at the publishing house for their professionalism, support, and guidance throughout the publication process. Your commitment to excellence has been instrumental in bringing this book to fruition.

We are indebted to DST-CPR at IIT Indore who have supported this project, providing resources and encouragement to pursue research in this important area of study. We extend our sincere gratitude to our family, friends, and colleagues for their persistent support and encouragement during this endeavor. Your encouragement and understanding have been invaluable in overcoming challenges and staying motivated. The authors acknowledge the Department of Science and Technology, Government of India, for funding the project entitled "Technological Innovation and Intellectual Property", DST/PRC/CPR/IITIndore (G).

Acknowledgments

Contents

Chapter 1
Basics of AI for Water Management

Abstract The chapter explores the fundamental principles and concepts related to incorporating Artificial Intelligence (AI) into water management. With growing global challenges in water resources, grasping the fundamentals of AI becomes vital for tackling complex issues. The chapter offers an overview of basic AI concepts, methodologies, and their relevance in the realm of water management. By examining essential components and considerations, this chapter acts as an introductory guide to understanding the central importance of AI in transforming approaches to sustainable and efficient water resource management. Moreover, the chapter goes beyond mere definitions and introduces readers to the practical applications of AI in delivering challenges related to water resources. By examining essential components and considerations, it offers valuable insights into how AI can be strategically employed to enhance the effectiveness and sustainability of water resource management practices.

1.1 Introduction to AI in Water Management

As the world's population grows, the need for clean water, an essential resource for humanity, also escalates. The Earth's population is expanding quickly, leading to a corresponding rise in the need for clean water. As per a report from the United Nations, the global demand for water is projected to surge by 50% by 2030. This places a significant burden on both our natural environment and the water sector, necessitating swift adaptation. It's crucial to take good care of our freshwater resources. While the worldwide demand for freshwater has risen significantly over the last century, the distribution of usage among different regions has shown remarkable stability. Organisation for Economic Co-operation and Development (OECD) nations have consistently maintained a usage share of approximately 20–25% of global freshwater resources. In contrast, Brazil, Russia, India, China, and South Africa (BRICS) countries have consistently held the position of the largest consumers, consistently accounting for about 45% of total usage. The world's freshwater resources are consistently used by 30–33% of the population, differing widely based on location, climate and a nation's reliance on agriculture or industry. Sustaining water resources requires usage below natural replenishment rates, including internal sources like rivers and rain-fed groundwater. Monitoring "renewable internal freshwater flows" is vital for assessing water security. Exceeding these flows signals declining resources. Per capita renewable freshwater resources, tied to total flows and population, can decrease due to lower rainfall or population growth. Many countries grapple with declining per capita resources due to population growth, presenting water challenges. Globally, agriculture utilizes approximately 70% of the world's freshwater resources. Low-income nations tend to allocate up to 90% of their water for agriculture, while middle-income countries allocate around 79%, and high-income countries allocate only 41%. Many countries in South Asia, Africa, and Latin America allocate over 90% of their water to agriculture, with Sudan leading at 96%. In contrast, industrial activities consume roughly 17% of total water withdrawals on a global scale, while municipal purposes account for approximately 12% (FAO and UN water, 2021; World Bank, 2019 data).

AI is transforming water resource management and environmental conservation. AI is increasingly applied to tackle water management complexities, from optimizing resource allocation to mitigating climate change effects. This overview explores AI's role in the water sector, emphasizing its potential to revolutionize decision-making, monitoring, and resource optimization. AI utilizes cutting-edge technologies such as machine learning, neural networks, natural language processing, and computer vision (Shrestha et al., 2023). These tools enable AI to analyze extensive datasets, recognize patterns, and offer predictions and recommendations based on data-driven insights. Machine learning (ML) has demonstrated its effectiveness as in incredibly versatile tool in the field of environmental science and engineering. While it can be complex to apply machine learning to water quality analysis and assessment, it has the potential to deliver more precise results, according to research by Zhu et al. (2022). Water comes in many different forms, including drinking water, wastewater, groundwater, surface water, seawater, and freshwater, each with its unique characteristics and research complexities. Past studies have shown that machine learning is a promising approach to address these challenges effectively.

AI systems can predict water contaminant composition, enhancing water treatment plant efficiency. Studies reveal that employing neural networks to optimize conventional controllers can lead to a significant 40% reduction in energy costs. Throughout history, human settlements have relied on accessible clean water sources. Due to the growing global population and diminishing freshwater quality, there is an ongoing search for technologies to secure a consistent and dependable supply of clean water. In recent times, machine learning has seen extensive utilization across diverse fields such as transportation, energy, healthcare, and manufacturing. Nevertheless, the integration of AI within the water sector continues to be constrained (Doorn, 2021).

However, AI has proven effective in modeling groundwater quality prediction (Hanoon et al., 2021; Kashani et al., 2023), forecasting water levels (Wee et al., 2021), optimizing irrigation practices (Gao et al., 2023; Patel et al., 2023; Akkem et al., 2023), promoting water conservation (Doorn, 2021), and enhancing water treatment processes (Safeer et al., 2022). The use of AI in water treatment is becoming increasingly popular as it addresses challenges in traditional methods. At present, the water sector is making substantial investments in AI, with forecasts suggesting an expected investment of 6.3 billion USD by 2030. The integration of AI into water treatment systems is anticipated to yield substantial cost reductions of 20 to 30% through the optimization of chemical usage and decreased operational expenditures. AI facilitates water treatment processes, simplifying their execution by its flexibility, generalization, and design simplicity, as highlighted by Alam et al. (2022). Multiple studies have shown the effectiveness of various AI tools in modeling and optimizing water treatment processes, including pollutant removal (Alam et al., 2022; Park et al., 2020; Tung and Yaseen, 2020).

An AI model has been implemented to keep an eye on groundwater levels in Brazilian aquifers by utilizing satellite gravimetry data, as per the study conducted by Camacho et al. (2023). The AI model's outputs were utilized to calculate the changes in groundwater storage in two significant aquifers over the past 20 years. The results revealed that water loss occurred as a result of an extended drought affecting a large portion of the country and an increase in groundwater extraction for irrigation purposes. The research suggests that employing AI techniques based on satellite data could offer an effective method for monitoring groundwater in areas where sophisticated monitoring infrastructure is lacking.

Recent studies have proven the value of utilizing AI to forecast groundwater quality (GWQ) across different aquifers. By using AI models with metaheuristic optimization, researchers can capture the nonlinearity of water quality parameters, improving the reliability and accuracy of predictions. Therefore, future research should focus on employing hybrid models to enhance the dependability of GWQ predictions further, as suggested by Hanoon et al. (2021). Furthermore, AI technologies have demonstrated notable progress in a range of water treatment procedures, including evaluating water quality, coagulation/flocculation, disinfection, membrane filtration, desalination, and simulating wastewater treatment facilities. Safeer et al. (2022) have highlighted the utilization of AI in predicting membrane fouling, removing heavy metals, and monitoring levels of biological oxygen demand (BOD) and chemical oxygen demand (COD). Additionally, Zhu et al. (2022) have emphasized the application of AI algorithms in evaluating water quality across different environments such as surface water, groundwater, drinking water, wastewater, and seawater. These advancements in AI hold promise to transform the water treatment sector, enhancing water quality for both human consumption and industrial purposes.

Earlier research has outlined four primary domains where AI is applied within the water industry: modeling, prediction, decision support, operational management, and optimization (Park et al., 2022; Doorn, 2021). To develop and apply responsible AI techniques, it is necessary for water professionals and data scientists to work collaboratively. Insights from social sciences and humanities are also essential for this purpose. AI is extensively used in predicting various adsorbents for removing dyes, metals, pharmaceuticals, drugs, organic compounds, pesticides, and PCPs from water (Alam et al., 2022). Although AI offers many advantages, its implementation in water treatment systems is limited by data availability, selection issues, and concerns about reproducibility.

Fu et al. (2022), performed an assessment of the present utilization of deep learning in addressing urban water challenges. The evaluation encompassed a range of aspects including forecasting water demand, identifying leaks and contamination, assessing sewer defects, predicting the state of wastewater systems, monitoring assets, and managing urban flooding. The study identified five crucial domains that require attention in future research in implementing deep learning for urban water management. These include data privacy, the development of algorithms, transparency, reliability, multi-agent systems, and digital twins. Additionally, Camacho et al. (2022) quantified changes in groundwater storage in the Urucuia and Bauru-Caiua aquifers during the previous 20 years. It was revealed that water loss occurred primarily as a result of prolonged drought affecting a significant portion of the country and an increased demand for groundwater due to intensified irrigation practices. Moreover, their research proposed that the integration of satellite data and AI presents a cost-effective solution for monitoring inadequately equipped aquifers on a continental scale, with the potential for global applicability. In recent studies, the concept of Adaptive Intelligent Dynamic Water Resource Planning (AIDWRP) is being adopted as a strategy for sustaining urban water environments. Within this approach, adaptive intelligence is a specific application of AI, effectively modeling environmental planning for sustainable water development. AI modeling can improve water efficiency by combining data-driven insights from AI tools with human skills (Xiang et al., 2021). During the last two decades, AI methods have seen extensive application in creating resilient models for handling various stochastic hydrological factors. These methods have demonstrated significant progress, particularly in formulating optimal guidelines for reservoir management. A comprehensive examination has explored the development of AI in predicting reservoir inflow and estimating evaporation from reservoirs, emphasizing their central importance in reservoir simulation (Allawi et al., 2018). The research found that AI models exhibit exceptional performance in managing data owing to their remarkable accuracy in handling non-linear data, resilience, dependability, affordability, problem-solving abilities, decision-making process, as well as efficiency and effectiveness. These AI models serve as excellent aids for overseeing, managing, promoting sustainability, and guiding policymaking concerning river water quality. The study provides an insight into the progress made in utilizing AI models to simulate river water quality between 2000 and 2020.

The incorporation of AI in water management has demonstrated its advantages. It encompasses data analysis, building regression models, and devising algorithms to establish effective water supply systems and networks. Additionally, this technology enables the assessment of water resource health and anticipation of probable challenges. Through the integration of AI into water management and infrastructure, we can cultivate intelligent water systems that are sustainable, economical, and flexible to evolving circumstances. This technological implementation optimizes water management solutions while anticipating potential problems (Mahardhika and Putriani, 2023).

1.2   Key AI Technologies for Water Solutions

In the opinion of water management, the adoption of AI technologies represents a revolutionary shift towards more efficient and sustainable solutions. Given the increasing difficulties presented by issues such as population growth, climate change, and urbanization, there is a pressing need for inventive solutions to manage water resources. AI, which includes a range of advanced innovations like machine learning, computer vision, and forecasting analytics, stands out as a valuable tool for tackling complex water-related problems. Leveraging extensive datasets, AI algorithms offer a wide range of functionalities, which are listed in Table 1.1 (Asian Development Bank, 2020).

The AI-based solutions listed in Table 1.2 are derived from the report "Artificial Intelligence Solutions for the Water Sector," addressing digital water concerns published by the International Water Association (2022).

According to Rechards et al. (2023), AI applications have not been widely implemented in real-world water systems. However, they believe that AI can provide significant benefits in three areas: (1) improving insights, catchment management, and emergency response in water supply at the catchment level; (2) making water distribution and disposal more efficient by aiding in the treatment, design, operation, and maintenance of network infrastructure at the network level; and (3) enhancing service availability, demand management, and water justice considerations for end-users. Figure 1.1 illustrates the potential advantages of employing AI to address challenges within water systems.

1.3   Case Studies in AI-Enabled Water Systems

Advancements in AI have revolutionized water systems management. This collection of case studies examines into the significant change of AI-enabled solutions on water systems, exploring real-world applications and their implications for sustainability, efficiency, and resource management. Through a diverse array of examples, we uncover how AI technologies are reshaping the landscape of water supply, distribution, and demand management. These case studies illuminate the innovative approaches, challenges, and successes that emerge when cutting-edge AI solutions intersect with the intricate dynamics of water systems. From optimizing network operations to proactive asset management, the stories within this compilation offer valuable insights into the practical implementation and outcomes of AI in the complex domain of water systems.

Table 1.1 Applications of AI for smart water management

1	Creating efficient configurations for monitoring and control networks	The digital evolution of water utilities involves monitoring and control networks, with AI algorithms strategically placing sensors to maximize system information while minimizing costs. This includes reducing control points and prioritizing cost-effective pressure gauges over pricier flowmeters. The quantitative approach aligns ICT investments with operational gains, offering a practical framework for often neglected cost–benefit analyses
2	Identifying numerical evidence of actual and water losses that are visible or noticeable	By employing state estimation and stochastic optimization, AI algorithms provide detailed information about water losses' quantity and nature in a given area. These algorithms continuously and probabilistically calibrate the network, departing from the traditional one-time calibration, allowing analysis of error structures at each control point. This approach enables insights into error patterns This method helps to distinguish various types of water losses. For instance, it can differentiate between pipe leaks and unauthorized consumption. The effectiveness of this approach relies on the density and frequency of measurements within particular network sectors. Although this numerical identification cannot replace the precise pinpointing of water leaks or field equipment connections, it offers a cost-effective alternative. By avoiding the deployment of leak detection teams, it saves time and resources. Additionally, it optimizes sectorization in distribution networks and prioritizes pipe replacement programs
3	Conservation of energy	Stochastic optimization techniques are essential for optimizing energy usage in network operations using AI algorithms. This optimization involves defining efficient operating procedures based on predetermined configurations and identifying economically viable investments, like pump replacements or changes in energy contracts, to achieve energy savings

(continued)

Table 1.1   (continued)

4	Developing backup plans and emergency procedures	Water utilities are well-equipped to handle emergency situations such as pipe bursts, equipment malfunctions, power outages, water scarcity, and contamination events. AI algorithms play a crucial role in optimizing responses by assessing risks, ranging from service disruptions to potential health hazards. By leveraging AI-driven techniques, contingency protocols can be predefined or determined in real-time to address issues like algal blooms or minimize consumer impact during a pipe burst. These optimizations are invaluable in ensuring swift and effective responses to unforeseen circumstances
5	Categorization of consumption patterns and forecasting demand	Water utilities prepare for emergencies to minimize the impact on customers by addressing issues such as pipe bursts, breakdowns, energy blackouts, water scarcity, and contamination. AI algorithms are essential for enhancing optimization of the response and assessing risks, ranging from service interruptions to health threats. Both predefined contingency protocols (such as addressing algal blooms) and real-time decisions (such as identifying valves during a pipe burst) can benefit from AI-driven optimization
6	Categorization of consumption patterns and demand forecasting	Using advanced statistical tools and historical data, AI algorithms continually refine their learning with additional information. This results in highly accurate real-time predictions of water demand at specific nodes or groups, spanning 24 h or even longer periods. These predictions facilitate long-term planning for capacity expansion, while incorporating uncertainty levels based on historical data that calibrates the hydraulic model. Additionally, long-term forecasts are tailored to match user-defined climatic and socioeconomic scenarios
7	Optimizing network expansion by designing an ideal configuration	Advanced AI optimization tools provide cost-effective configurations that take into account both capital and operational expenditures, as well as specific targets. Specialized AI algorithms identify optimal alternatives for network expansion, considering uncertainties such as population forecasts and spatial urban growth. This approach enhances decision-making with a more resilient strategy
8	Programs for actively managing assets	Water utilities prioritize maintenance and replacement activities to provide high-quality service while keeping costs low. They use active asset management to make proactive decisions instead of reacting to external events. To determine the best schedules for asset monitoring and replacement, algorithms analyze statistical factors, such as useful life, criticality, and other variables

Table 1.2 The usage of artificial intelligence in the water industry to provide efficient, effective, and sustainable solutions

1	Real-time Detection of Pipe Bursts in Water Distribution Networks	The innovative AI-driven solution addresses the global challenge of water distribution systems by detecting pipe bursts, leaks and equipment failures within the infrastructure. Operating in near real-time, the system analyzes signals from pressure and flow sensors using Artificial Neural Networks to predict future values. The system detailed below utilizes predictions and real-time observations to gather evidence of potential failure events. These events are then analyzed using Bayesian Networks to estimate the probability of their occurrence and to activate alarms as needed. Originally developed as part of a research initiative, this system has since been integrated into a commercial Event Detection System (EDS). It has been in active use by a prominent UK water company since 2015, processing data from over 7000 sensors every 15 min. This ensures both timely and accurate detection of pipe bursts and leaks, with a minimal rate of false alarms. Additionally, EDS not only detects incidents but also takes proactive measures to prevent burst events by identifying equipment failures, such as pressure reducing valve malfunctions, without the need for a hydraulic or simulation model of the water distribution network. EDS's triumph has led to noteworthy reductions in operational expenses, fewer instances of customer supply downtime, decreased leakage, and an enhanced overall service provision for its expansive customer base of over 7 million. Although specific details are not disclosed for commercial reasons, the impact of EDS has positively influenced the company's business culture
2	AI and Computer Vision for Automated Asset Condition Assessment	Detecting structural faults in urban drainage (sewer) systems is crucial to prevent pollution and flooding. Traditional manual analysis of CCTV videos is time-consuming and subjective. An AI-based solution that uses computer vision and machine learning has been developed to automate CCTV video analysis, achieving high accuracy in fault detection. The technology has been successfully tested with real data from various water companies and is now being commercialized in partnership with a UK water company
3	Preventive Control of Wastewater Treatment Plants	Royal Haskoning DHV's Aquasuite® software, deployed at PUB Singapore's Integrated Validation Plant, boosts plant efficiency by providing predictive insights. Aquasuite PURE collects up-to-the-minute information on fluid movements and quality assessments, utilizing sophisticated analytics and machine learning to predict wastewater characteristics. It optimizes treatment processes in real-time, achieving an 88% influent flow prediction accuracy. Interfacing with the SCADA system, it transmits data to the Aquasuite cloud, providing advanced analytics and maintaining on-premises optimization. Machine learning evaluates process efficiency, enabling Aquasuite to learn and predict operations, functioning as an autopilot. Initial results indicate a 15% reduction in aeration flow with predictive control, leading to energy savings

(continued)

Table 1.2 (continued)

4	Intelligent Alerts for Proactive Management of Wastewater Networks	Aquasuite FLOW is a predictive analytics tool that uses AI to forecast high-water levels, detect pollution events, and identify anomalies that signal blockages in wastewater networks. During rainfall, the risk of blockages rises due to factors like flooding and foreign objects. Aquasuite uses AI to identify anomalies in sewer flows, minimizing false alarms with a customizable alarm system. Monitors collect data transmitted to the Aquasuite cloud for analysis, integrating two multi-layer perceptron ANNs for predictive flow levels and anomaly identification. Continuous machine learning improves anomaly detection, enabling proactive customer notifications, spill prevention, and timely blockage clearance
5	Proactive Asset Management using Bayesian Networks	A probabilistic graphical model with Bayesian inference is used to predict vulnerable pipes in global pilot projects in Stockholm, Singapore, the UK, and Denmark
6	Rainfall Monitoring using Computer Vision	With the Internet of Things, unconventional sensing, including video feeds from smartphones or CCTV cameras, improves spatiotemporal representation. Various technologies such as image processing, computer vision, and AI are being used to estimate rainfall intensity. Companies like WaterView and the Hydroinformatics Institute have developed realistic methods to address meteorological risks in different areas. Universities such as the Southern University of Science and Technology in China are also contributing to this field

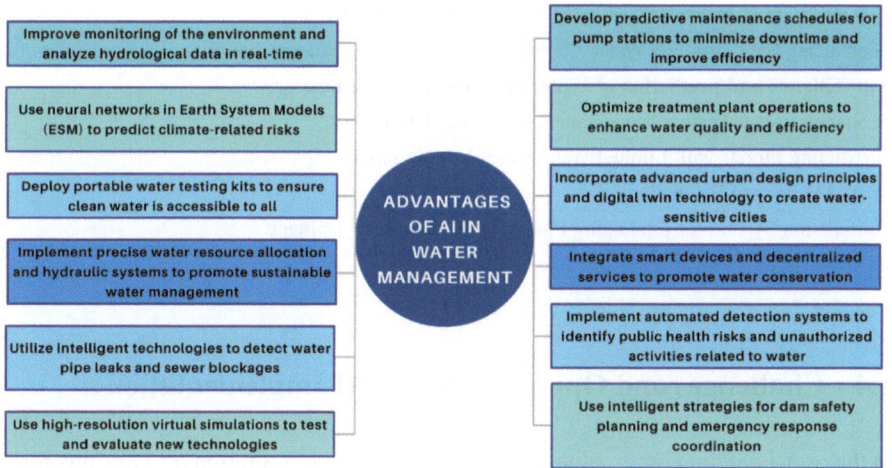

Fig. 1.1 The prospective advantages of utilizing AI in addressing challenges within water systems

Monitoring of Future Water Infrastructures (Raconteur, 2016)
AI-driven water supply infrastructure in Singapore, Asia
Virtual water network – HydroIQ, Kenya, Africa
Assessing the state and potential hazards of drinking water pipelines across the United States
AI-enabled sewer monitoring in the USA
Forecasting Water Demand and Consumption
Predicting district-level water demand in Spain
"Smart Water Grid" is an urban water supply and management system that is being implemented in Australia
Modular modelling-based forecasting in London, UK
Multi-level Prediction in Southern California, USA
Observing Water Reservoirs and Dams
Identifying High-Hazard Potential Dams in the United States
Mitigating Greenhouse Gas Emissions from Amazonian Dam Projects
Continuous Monitoring of Dams in Real-Time, United Kingdom
Smart Dam Monitoring in Italy
Preserving Lake Sulunga using the Africa Regional Data Cube (ARDC) in Tanzania
Assessing Water Quality
Clean Water AI
Forecasting Water Quality in Iran (Talesh et al., 2019)
Forecasting Groundwater Quality in India (Kadam et al., 2019)
Enhancing Safe Drinking Water Access Through Satellite Data in Ghana
Surveillance and prediction of Water-Related Catastrophes
Enhancing Wet Weather Management with AI in Ohio, USA
Predicting Floods with Limited Water Level Data in Japan
Rapid Hurricane Evaluation
Simulation System for Evaluating Flood-Induced Industry Damage

1.4   Challenges and Opportunities in AI Implementation

In the ever-changing realm of technology today, the incorporation of AI brings forth numerous challenges and possibilities across diverse sectors. As AI continues to evolve, it introduces transformative potential, reshaping the way industries operate and enhancing efficiency in unprecedented ways. This duality of challenges and opportunities necessitates a comprehensive exploration to navigate the complexities associated with AI implementation. From ethical considerations and regulatory frameworks to the potential for innovation and enhanced decision-making, the spectrum of challenges and opportunities in AI implementation demands careful examination. This exploration aims to unravel the intricate tapestry of AI's potential pitfalls and benefits, shedding light on the key considerations that accompany the integration of this powerful technology into diverse domains.

Realizing the complete potential of AI for environmental advantages involves tackling issues related to governance, resourcing, deployment, collaboration, and data maturity. To effectively address environmental concerns such as climate change, biodiversity loss, and pollution, multidisciplinary collaboration is imperative among governments, the private sector, technology companies, industries, NGOs, and academia. The swift progress of AI offers a distinctive chance to deploy robust and accurate tools in addressing these challenges, underscoring the importance of collective action for the well-being of the planet.

1.4.1 Challenges and Opportunities in Water Management

Challenges

From the standpoint of data quality and accessibility, a prominent hurdle lies in the fragmented, inconsistent, and outdated nature of water-related data, posing challenges in effectively training AI models. Moreover, the complexity inherent in water systems, influenced by diverse factors such as climate dynamics, geographical variations, and human behaviors, presents formidable challenges for AI algorithms seeking to provide meaningful insights and solutions. Effective water management necessitates interdisciplinary collaboration across hydrology, engineering, and policy domains, requiring concerted efforts to align objectives and communication channels. Ethical considerations surrounding data privacy, equity, and accountability loom large, demanding transparent decision-making processes to instill trust in AI technologies. Furthermore, navigating the sophisticated regulatory and legal frameworks governing water management across regions and jurisdictions poses additional obstacles to AI implementation.

Opportunities

Despite the challenges, AI offers immense potential to revolutionize water management practices. By harnessing AI's capabilities for data-driven decision-making, water managers can glean valuable insights from vast datasets to inform more informed decision-making processes. Predictive analytics powered by AI algorithms enable the forecasting of water availability, demand, and quality, facilitating proactive planning and resource allocation strategies. Furthermore, AI holds promise in optimizing the operation of water infrastructure systems, enhancing efficiency and cost-effectiveness. Early warning systems fueled by AI can detect and predict water-related hazards like floods, droughts, and pollution events, enabling timely response and mitigation efforts. Additionally, AI technologies can foster community engagement and participation in water management initiatives through innovative platforms for data visualization, citizen science, and participatory decision-making.

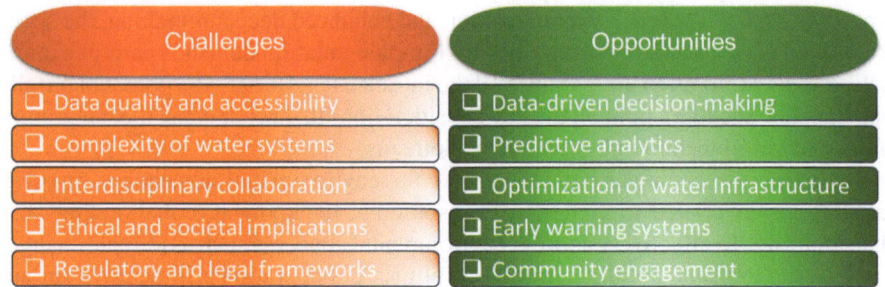

Fig. 1.2 Challenges and Opportunities in AI Implementation for Water Management

Figure 1.2 illustrates the Challenges and Opportunities in AI Implementation for Water Management.

AI-Powered Solutions for Water Assessment

- Showcase commercial viability.
- Establish trust in technology.
- Integrate participatory approaches.
- Increase access to data.

The ethical aspects of utilizing AI in water-related fields are frequently disregarded. Recent discussions on AI ethics suggest five fundamental principles that should be taken into account: transparency, fairness and equity, accountability, privacy, and avoiding harm (Doorn, 2021). Researchers emphasize the significance of prioritizing the 'responsible implementation' of AI in water systems to alleviate potential risks. This applies to scholars, professionals in both the water and AI sectors, and ordinary users. The authors put forward three key suggestions to guarantee the safe and accountable integration of AI in water systems. These recommendations involve filling in the deficiencies in basic infrastructure and digital literacy, setting up strong institutional, software, and hardware mechanisms for reliable AI, and giving precedence to applications through a thorough assessment of benefits and risks (Richards et al., 2023).

Issues related to infrastructure, human resources, direct threats such as design flaws and misuse, and indirect dangers stemming from interconnected system breakdowns could diminish the advantages of AI if not handled with care (Richards et al., 2023). Figure 1.3 depicts the possible threats posed by AI in water systems, which could result in developmental hurdles. It's crucial to handle these challenges prudently to safeguard the full potential of AI benefits. These challenges encompass concerns regarding infrastructure, human resources, as well as direct risks such as design flaws and misapplication, alongside indirect risks stopping from systemic breakdowns. A research conducted by Richards et al. (2023) underlines these potential hazards, as illustrated in Fig. 1.3.

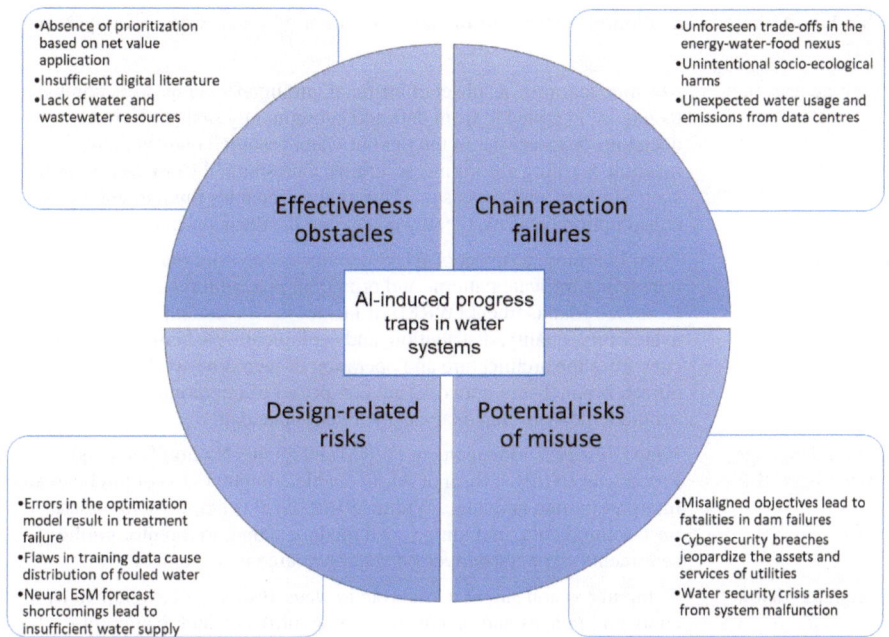

Fig. 1.3 Potential risks associated with AI in water systems that may lead to developmental challenges

1.5 Future Trends and Innovations in AI for Water Sustainability

As our global community grapples with the escalating issues of water scarcity, pollution, and efficient resource management, the integration of AI emerges as a crucial force shaping the future of water sustainability. In recent years, the merging of AI and water management has seen significant advancements, presenting unparalleled opportunities to tackle longstanding challenges and envision a more resilient, intelligent, and sustainable water infrastructure.

This section investigates the forefront of Future Trends and Innovations in AI for Water Sustainability, exploring the dynamic landscape where cutting-edge technologies converge with the imperatives of environmental conservation and water resource optimization. From precision monitoring and predictive analytics to proactive management strategies, AI's role in water sustainability spans a spectrum of applications. Table 1.3 shows various AI-based technologies relevant to the administration of water assets (Kamyab et al., 2023).

Advancements in Big Data Analytics (BDA) are instrumental in revolutionizing water resource management (WRM) by offering essential tools for comprehensive analysis and well-informed decision-making. BDA introduces numerous innovative applications in the water domain:

Table 1.3 AI-based technologies relevant to the administration of water assets (Kamyab et al., 2023)

Machine learning	Machine learning, a subset of artificial intelligence, empowers computers to acquire knowledge from data and subsequently make informed decisions or forecasts. In the field of water resource management, machine learning algorithms scrutinize data sourced from various outlets such as water quality sensors. This analysis enables them to anticipate consumption patterns, thereby enhancing the decision-making framework
Deep learning	Deep Learning, a subset of AI, is demonstrating remarkable utility in unraveling intricate patterns and providing precise forecasts for Water Resource Management (WRM). It tackles issues encompassing water availability, quality, distribution, and ecological well-being by taking cues from the architecture and operation of neural networks found in the human brain. These networks are comprised of interconnected layers of synthetic neurons that analyze and manipulate data
Natural language processing (NLP)	Water Resource Management (WRM) integrates Natural Language Processing (NLP), a form of AI, to enable computers to comprehend and interpret human language. Within WRM, NLP is employed to assess unstructured data, including social media content, to monitor public sentiments and perspectives on water resource management
Computer vision	Computer vision shows a vital role in water resource management by analyzing images and videos to monitor shifts in water availability, consumption, and quality. This technology processes various visual data sources like satellite imagery, offering valuable insights into water resource dynamics
Predictive analytics	Predictive analytics, a branch of AI, employs machine learning algorithms to examine data for predicting events in water resource management, such as patterns in water usage during droughts, trends in water quality, and other crucial indicators

- BDA facilitates real-time monitoring of water resources, enabling more effective water management decisions by monitoring changes in water quality, distribution, and demand.
- Optimizing Water Allocation: Through the analysis of water usage patterns and the utilization of diverse data sources, BDA optimizes water allocation to ensure efficient and fair distribution of water resources, addressing key concerns in water management.
- Improving Water Demand Forecasting: Accurate prediction of water demand is essential for effective water management. BDA's predictive capabilities utilize data from water infrastructure, usage patterns, and various sources to forecast water demand with precision, enabling decision-makers to ensure a sustainable and equitable water supply, which is vital for effective water management.
- Detecting Wastewater: BDA aids in identifying leaks and inefficiencies in water infrastructure, providing insights to minimize wastewater and optimize water usage patterns, contributing to the adoption of more sustainable water management practices.

- Enhancing Water Quality Monitoring: BDA enhances water quality monitoring by enabling the identification of patterns and predictions regarding water parameters, facilitating informed decision-making for maintaining water quality standards.

Funding This study was funded by the Department of Science and Technology, Government of India [DST/PRC/CPR/IITIndore (G)] for the project entitled "Technological Innovation and Intellectual Property"

References

Akkem, Y., Biswas, S.K., Varanasi, A. (2023). Smart farming using artificial intelligence: A review. *Eng. Appl. Artif. Intell. 120*, 105899.

Alam, G., Ihsanullah, I., Naushad, M., & Sillanpää, M. (2022). Applications of artificial intelligence in water treatment for optimization and automation of adsorption processes: Recent advances and prospects. *Chemical Engineering Journal, 427*, 130011.

Allawi, M. F., Jaafar, O., Mohamad Hamzah, F., Abdullah, S. M. S., & El-Shafie, A. (2018). Review on applications of artificial intelligence methods for dam and reservoir-hydro-environment models. *Environmental Science and Pollution Research, 25*, 13446–13469.

Asian Development Bank (2020). Jenny, H., Alonso, E. G., Wang, Y., & Minguez, R. (2020). Using artificial intelligence for smart water management systems.

Camacho, C.R., Getirana, A., Rotunno Filho, O.C., Mourão, M.A.A. (2022). Large-scale groundwater monitoring in Brazil assisted with satellite-based artificial intelligence techniques. *Authorea Prepr.*

Chhipi-Shrestha, G., Mian, H. R., Mohammadiun, S., Rodriguez, M., Hewage, K., & Sadiq, R. (2023). Digital water: artificial intelligence and soft computing applications for drinking water quality assessment. *Clean Technologies and Environmental Policy*, 1–30.

Connecting the Drops: Global Water Security and Sanitation Partnership Annual Report 2019 (English). Umbrella Trust Fund Annual Report Washington, D.C.: World Bank Group. http://documents.worldbank.org/curated/en/997021571169834156/Connecting-the-Drops-Global-Water-Securityand-Sanitation-Partnership-Annual-Report-2019

Doorn, N. (2021). Artificial intelligence in the water domain: Opportunities for responsible use. *Science of the Total Environment, 755*, 142561.

FAO and UN Water. 2021. Progress on Level of Water Stress. Global status and acceleration needs for SDG

Fu, G., Jin, Y., Sun, S., Yuan, Z., & Butler, D. (2022). The role of deep learning in urban water management: A critical review. *Water Research*, 118973.

Gao, H., Zhangzhong, L., Zheng, W., Chen, G. (2023). How can agricultural water production be promoted? a review on machine learning for irrigation. *J. Clean. Prod.* 137687.

Hanoon, M. S., Ahmed, A. N., Fai, C. M., Birima, A. H., Razzaq, A., Sherif, M., ... & El-Shafie, A. (2021). Application of artificial intelligence models for modeling water quality in groundwater: comprehensive review, evaluation and future trends. *Water, Air, & Soil Pollution, 232*, 1–41.

Hasanpour Kashani, M., Nikpour, M. R., & Jalali, R. (2023). Water quality prediction using data-driven models case study: Ardabil plain, Iran. *Soft Computing, 27*(11), 7439–7448.

https://doi.org/10.1007/s00500-022-07684-7

https://blogs.worldbank.org/water/future-water-how-innovations-will-advance-water-sustainability-and-resilience-worldwide

https://blogs.worldbank.org/water/tapping-water-innovation-new-partnership-accelerate-access-water-and-wastewater-technologies

https://blogs.worldbank.org/water/water-cant-wait-accelerating-adoption-innovations-water-security

https://ourworldindata.org/water-use-stress

https://www.epa.gov/water-innovation-tech/examples-innovation-water-sector

International Water Association (2022). Digital Water Artificial Intelligence Solutions for the Water Sector. https://iwa-network.org/wp-content/uploads/2020/08/IWA_2020_Artificial_Int elligence_SCREEN.pdf

Kadam, A. K., Wagh, V. M., Muley, A. A., Umrikar, B. N., & Sankhua, R. N. (2019). Prediction of water quality index using artificial neural network and multiple linear regression modelling approach in Shivganga River basin, India. *Modeling Earth Systems and Environment, 5*, 951–962.

Kamyab-Talesh, F., Mousavi, S. F., Khaledian, M., Yousefi-Falakdehi, O., & Norouzi-Masir, M. (2019). Prediction of water quality index by support vector machine: A case study in the Sefidrud Basin, Northern Iran. *Water Resources, 46*, 112–116.

Kamyab, H., Khademi, T., Chelliapan, S., SaberiKamarposhti, M., Rezania, S., Yusuf, M., Fara-jnezhad, M., Abbas, M., Jeon, B.H., Ahn, Y. (2023). The latest innovative avenues for the utiliza-tion of artificial Intelligence and big data analytics in water resource management. *Results Eng.* 101566.

Mahardhika, S. P., & Putriani, O. (2023, June). Deployment and use of Artificial Intelligence (AI) in water resources and water management. In *IOP Conference Series: Earth and Environmental Science* (Vol. 1195, No. 1, p. 012056). IOP Publishing.

Park, J., Lee, W. H., Kim, K. T., Park, C. Y., Lee, S., & Heo, T. Y. (2022). Interpretation of ensemble learning to predict water quality using explainable artificial intelligence. *Science of the Total Environment, 832*, 155070.

Patel, A., Kethavath, A., Kushwaha, N. L., Naorem, A., Jagadale, M., Sheetal, K. R., & Renjith, P. S. (2023). Review of artificial intelligence and internet of things technologies in land and water management research during 1991–2021: A bibliometric analysis. *Engineering Applications of Artificial Intelligence, 123*, 106335.

Raconteur. (2016). Future of Water. 424, 16. Retrieved from http://rcnt.eu/wn0

Renna Camacho, C., Getirana, A., Rotunno Filho, O. C., & Mourão, M. A. A. (2023). Large-Scale Groundwater Monitoring in Brazil Assisted With Satellite-Based Artificial Intelligence Techniques. *Water Resources Research, 59*(9), e2022WR033588.

Richards, C. E., Tzachor, A., Avin, S., & Fenner, R. (2023). Rewards, risks and responsible deployment of artificial intelligence in water systems. *Nature Water,* 1–11.

Safeer, S., Pandey, R. P., Rehman, B., Safdar, T., Ahmad, I., Hasan, S. W., & Ullah, A. (2022). A review of artificial intelligence in water purification and wastewater treatment: Recent advancements. *Journal of Water Process Engineering, 49*, 102974.

Tung, T. M., & Yaseen, Z. M. (2020). A survey on river water quality modelling using artificial intelligence models: 2000–2020. *Journal of Hydrology, 585*, 124670.

Wee, W. J., Zaini, N. A. B., Ahmed, A. N., & El-Shafie, A. (2021). A review of models for water level forecasting based on machine learning. *Earth Science Informatics, 14*, 1707–1728.

Xiang, X., Li, Q., Khan, S., & Khalaf, O. I. (2021). Urban water resource management for sustainable environment planning using artificial intelligence techniques. *Environmental Impact Assessment Review, 86*, 106515.

Zhu, M., Wang, J., Yang, X., Zhang, Y., Zhang, L., Ren, H., ... & Ye, L. (2022). A review of the application of machine learning in water quality evaluation. *Eco-Environment & Health.*

Chapter 2
AI for Water Conservation

Abstract The chapter investigates the application of AI in the domain of water conservation. With escalating concerns about water scarcity and the need for sustainable resource management, AI emerges as a promising technology for optimizing water usage. The abstract provides insights into various AI-driven methodologies employed in water conservation efforts, ranging from smart water metering to predictive analytics. It explores how AI enhances efficiency in monitoring, managing, and conserving water resources. By examining case studies and highlighting innovative approaches, this exploration contributes to the ongoing discourse on leveraging AI to address critical challenges in water conservation, paving the way for more intelligent and sustainable water management practices.

2.1 Introduction to AI in Water Conservation

The increasing demands of a growing global population and the impacts of climate change have intensified the need for effective water conservation. In response to these challenges, technology has emerged as a crucial ally, with Artificial Intelligence (AI) standing out as one of the most promising innovations. This introduction delves into the convergence of AI and water conservation, illuminating its potential to revolutionize the management and preservation of this vital resource. Amid the complexities of modern-day water challenges, traditional conservation methods often fall short in promoting environmental sustainability. The necessity for advanced, data-driven solutions has become more pronounced than ever. Herein lies Artificial Intelligence, a transformative force capable of processing extensive data, making informed decisions, and optimizing processes with unparalleled efficiency. AI's role in water conservation extends beyond conventional methods, providing a dynamic and adaptive solution to the intricate problems faced by global water management systems. From smart water distribution to precision irrigation, the applications of AI are diverse and promising. In the subsequent sections, we will explore further the diverse aspects of AI in water conservation, exploring its applications, benefits,

M. K. Goyal et al., *AI Innovation for Water Policy and Sustainability*,
SpringerBriefs in Water Science and Technology,
https://doi.org/10.1007/978-3-031-72014-7_2

challenges, and the pivotal role it plays in shaping a sustainable future for the most precious resource.

Scientists are utilizing advanced technologies such as AI to transform water resource management on a global scale, fostering sustainable water supplies. AI, often hailed as the pinnacle of technology, introduces the concept of 'digital water' through continuous and adaptive processing of extensive datasets. This tech-driven approach empowers water managers and governmental bodies to establish efficient water systems, effectively addressing challenges related to water volatility and sustainability. Key tasks for water management involve sourcing new water reservoirs and ensuring the sustainable management of existing reserves. AI-driven algorithms and programs are under exploration for creating water plants that offer real-time data and contribute to the development of future models. Software-driven systems with sensors and neural networks dynamically plan water operations, integrating predictive models, robotic sensors, and blockchains for financial transactions. In a significant experiment, researchers from the Key Laboratory of Eco-hydrology of the Inland River Basin used AI to predict groundwater reserves, showing the potential of hybrid models for enhancing global water systems. In a significant experiment, scientists from the primary research facility focusing on Eco-hydrology in the Inland River Basin used AI to predict groundwater reserves, showing the potential of hybrid models for enhancing global water systems. Collaborative efforts among scientists and analysts on a global scale are imperative to contribute to the sustainable utilization of this invaluable resource. Leveraging AI in water resource management has demonstrated its ability to facilitate efficient water distribution and promote water conservation systems (Mahardhika and Putriani, 2023). Figure 2.1 demonstrates AI's role in water management for enhancing economic distribution and conservation systems.

2.2   The Role of AI in Enhancing Water Use Efficiency

In the face of escalating challenges posed by global population growth, climate change, and the unsustainable consumption of water resources, the quest for innovative solutions has never been more critical. AI stands at the forefront of the effort to enhance water use efficiency. As water scarcity becomes an increasingly urgent concern, the role of AI in optimizing water management practices takes center stage. This section explores the transformative impact of AI on water use efficiency, examining its applications across various sectors and its potential to reshape our approach to water conservation for a sustainable future. Figure 2.2 depicts the impact and applications of AI in enhancing water use efficiency.

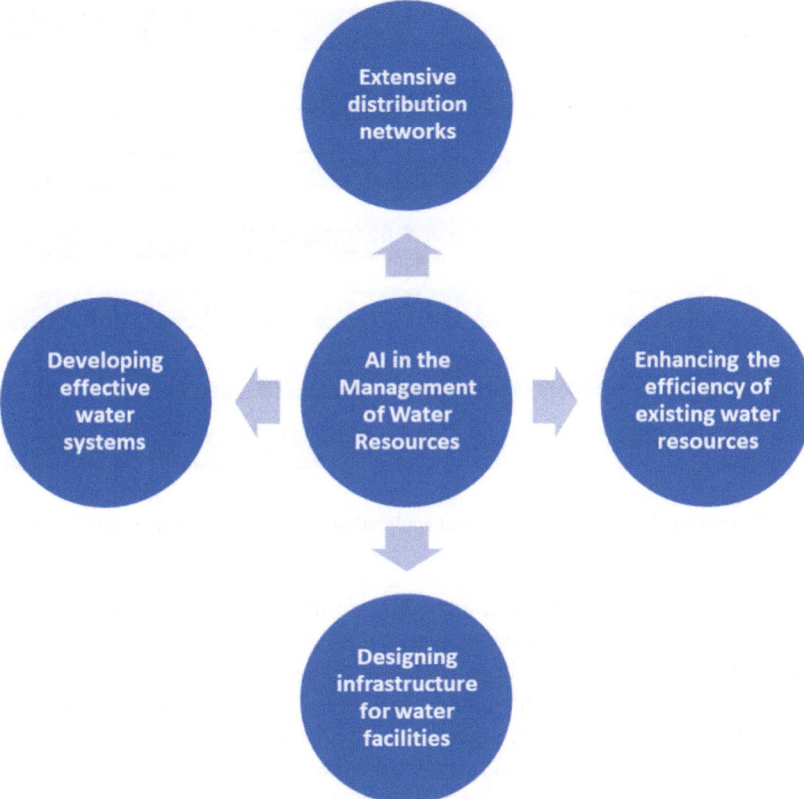

Fig. 2.1 AI's role in water management for enhancing economic distribution and conservation systems

2.2.1 Precision in Irrigation Practices

The FAO's 2021 report on the State of the World's Land and Water Resources for Food and Agriculture highlights the critical state of global agricultural systems, which are reaching their limits and having a significant impact on the world's food supply. As the demand for food security rises and water scarcity grows, there is a pressing need to sustain agricultural growth without causing additional harm to the environment. The intricate and nonlinear nature of factors influencing crop water consumption, including environmental conditions, soil quality, and crop type, complicates effective irrigation management. Traditional methods are evolving, with a shift towards incorporating automated and intelligent modern management approaches, particularly through the integration of AI technology. This adoption of AI in irrigation systems aims to address inefficiencies and significant water wastage associated with conventional methods. The use of AI-based models allows for the identification

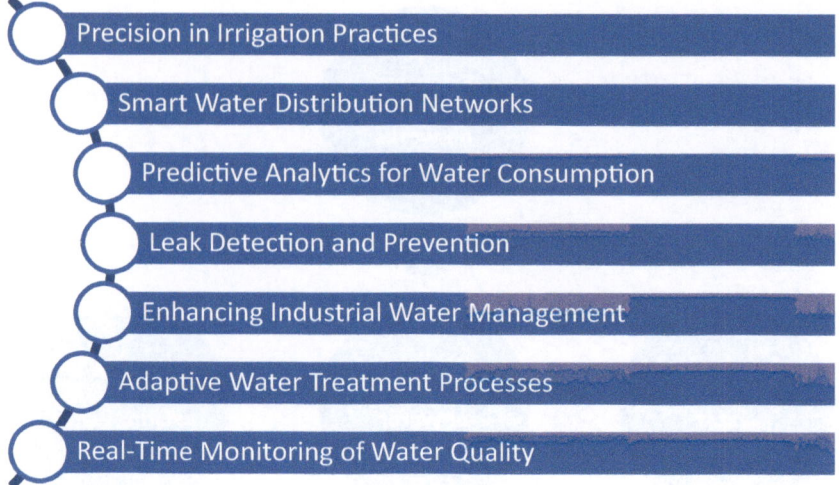

Fig. 2.2 Illustration depicting the impact and applications of AI in enhancing water use efficiency

and selection of inputs, enabling a more precise mapping of relationships between variables. Over the past two decades, various AI models have been applied across engineering disciplines to tackle scientific challenges, especially those involving complex, nonlinear relationships among variables. The emergence of smart irrigation, utilizing machine learning algorithms, proves to be an effective solution for conserving water in both parks and farmland under challenging weather conditions. While current irrigation systems have reduced water usage, smart irrigation addresses issues of uneven watering, over- or under-irrigation, and plant stress. Recent studies have explored sensor technologies capable of forecasting irrigation requirements by analyzing soil moisture levels at varying depths. In the irrigation industry, three types of smart irrigation systems are prevalent: soil moisture-based, weather-based, and hybrid-based. These systems employ machine learning algorithms to predict watering requirements by taking into account various factors such as plant type, growth stage, weather conditions, infiltration rate, and soil type. The integration of AI in agriculture and irrigation represents a promising avenue for sustainable resource management, ensuring the balance between agricultural productivity and environmental conservation.

2.2.2 Smart Water Distribution Networks

AI-driven smart water distribution systems are becoming instrumental in managing urban water supplies. These systems continuously monitor water flow, detect leaks, and dynamically adjust distribution based on demand. This real-time optimization minimizes losses, enhances efficiency, and ensures a reliable water supply for growing urban populations.

2.2.3   Predictive Analytics for Water Consumption

Harnessing the power of data analytics, AI enables accurate predictions of water consumption patterns. By analyzing historical data and considering variables like population growth and climate trends, AI models forecast future water demands. This foresight allows authorities to proactively plan and allocate resources, preventing shortages and ensuring sustainable water management.

2.2.4   Leak Detection and Prevention

Water leaks are a major factor in wasting resources. AI has become essential in detecting and preventing leaks by constantly monitoring water infrastructure. Utilizing machine learning algorithms, AI systems can identify abnormalities in water flow, allowing for quick intervention and reducing water waste.

2.2.5   Enhancing Industrial Water Management

Industries are major consumers of water, and AI offers innovative solutions to improve their water use efficiency. AI-driven systems optimize industrial processes, identifying opportunities to reduce water consumption without compromising productivity. Not only does this contribute to cost savings for businesses, but it also benefits the environment.

2.2.6   Adaptive Water Treatment Processes

AI aids in optimizing water treatment processes by adapting to changing conditions in real-time. AI-powered smart systems can adjust water treatment parameters based on specific requirements and variations in water quality, resulting in efficient and customized water treatment.

2.2.7 Monitoring Water Quality in Real-Time

Ensuring the quality of water is as crucial as managing its quantity. AI enables the monitoring of water quality in real-time by analyzing various parameters, including chemical composition and contamination levels. Rapid detection of anomalies allows for prompt interventions, safeguarding water quality for consumption and ecosystem health.

2.2.8 Personalized Water Consumption Feedback

At the individual level, AI applications can provide personalized feedback on water consumption. Smart home devices equipped with AI algorithms analyze usage patterns, offering insights and recommendations to users on how to reduce their water footprint. This personalized approach encourages responsible water use at the grassroots level.

In brief, the incorporation of AI in water management signifies a transformative leap towards enhancing water use efficiency. From agriculture to urban planning, industries, and individual households, AI's applications are diverse and impactful. As we navigate the challenges of a water-scarce world, the role of AI in optimizing water utilization becomes not just significant but imperative towards a future that is sustainable and water-secure.

2.3 Predictive Modeling for Water Conservation

Predictive modeling, leveraging advanced technologies and data analytics, presents an innovative approach to predict, understand, and address the intricate dynamics of water availability, usage, and conservation. This paradigm shift in water management holds the promise of optimizing resources, mitigating risks, and fostering a more sustainable and resilient water ecosystem. Table 2.1 shows the role of Microsoft's AI for Earth in ensuring future water conservation.

2.4 Real-Time Monitoring of Water Resources Using AI

Despite the current limited real-world implementation, we emphasize the expected benefits of AI applications in three key areas of water systems. These include enhancements in water supply management at the catchment level, optimization of water distribution and disposal at the network level, and improvements in meeting water demand at the end-user level.

Table 2.1   Microsoft's AI for Earth highlights the pivotal role of AI in securing the future of water conservation

Projects name	Uses
Observing the provision of potable water	Scholars at Stanford University's Natural Capital Project are applying machine learning techniques and remote-sensing data to identify smaller dams and reservoirs and providing the algorithm freely to the sustainable development community with the help of Microsoft Azure
Water Management in Megacities	The Indian Institute of Science in Bangalore, through its Department of Computational and Data Science, is harnessing the potential of the Internet of Things (IoT) technology to address the challenge of meeting the increasing water demand in India. Focusing on densely populated areas and megacities grappling with water scarcity, it employs data analytics and machine learning in the EqWater project. This approach optimizes water management through enhanced scheduling, leak detection, and predictive analysis of reservoir flow, seasonal weather, and residential usage to address inequitable water distribution and potential shortages
Improved understanding of weather	The Scripps Institution of Oceanography at UC San Diego hosts the Center for Western Weather and Water Extremes, which is dedicated to enhance its understanding of lesser-known weather phenomena in the western United States. One of these occurrences is atmospheric rivers, characterized by substantial masses of water vapor in the atmosphere. Despite limited understanding, they play a significant role in causing storms, floods, and influencing water supplies. The team employs deep learning to predict the behavior of these atmospheric rivers
Exploring the Impact of Tree Depletion	The western U.S. is witnessing a significant loss of trees due to drought, climate change, wildfires, and insect infestation, raising concerns about forest health. Tony Chang and his team at Conservation Science Partners leverage cloud computing and machine learning techniques to analyze imagery sourced from NASA and other platforms. Their focus is on evaluating tree health and estimating biomass. By connecting this data with regional water source information, they unveil intricate links between forest conservation, management practices, and water supplies
Forecasting Harmful Algae Blooms	Africa Flores, a research scientist at the University of Alabama, heads a team utilizing artificial intelligence to forecast detrimental algal blooms. These blooms, which can be dangerous to both humans and wildlife, are being analyzed by Flores' team by utilizing satellite images and weather models. The team is concentrating on Lake Atitlán in the Guatemalan Highlands. Flores employs machine learning to examine diverse factors like rainfall, temperature, and cloud cover, aiming to comprehend the conditions preceding algal blooms more thoroughly. This understanding could potentially lead to preventive measures being taken and improve agricultural practices. Flores intends to expand her research to encompass additional freshwater bodies across Central and South America, aiming to enhance its relevance and applicability on a broader scale

2.4.1   Water Supply Management at the Catchment Level

Machine learning models adeptly manage large datasets, including interferometric synthetic aperture radar imagery, effectively reconstructing missing data. This enables precise quantitative evaluations of historical freshwater dynamics. Moreover, complementary algorithms analyze diverse data streams, contributing to real-time monitoring and swift forecasts of hydrological cycles and weather events. These algorithms encompass both quantitative and qualitative factors.

AI applications optimize aquifer drawdown and dam-filling schedules, prevent harm to aquatic ecosystems, and automate detection of public health hazards and illegal activities in water bodies. In emergency prevention and response, artificial intelligence collaborates with up-to-the-minute rainfall data and early warning systems to oversee reservoir inflows and coordinate safe spillway releases. Intelligent groundwater management enhances resilience in areas susceptible to drought.

Neural Earth system models enhance our understanding of underlying physics and expand simulation options. Optimization algorithms play a crucial role in aiding the long-term planning of catchment watersheds and infrastructure. AI-enabled ESM outputs examine climate risks efficiently and inform the expansion of artificial water sources in areas anticipating water scarcity. AI-enhanced hydraulic models refine river engineering, improvements to dams and the implementation of storm surge barriers.

2.4.2   Enhancing the Management of Water Distribution and Drainage Systems at the Network Level

Given the increasing demands on water systems due to population growth, it is essential to leverage AI to develop new infrastructure for potable water, stormwater, and sewerage. This is necessary to address the challenges posed by aging critical assets. AI systems, when aligned with specific goals and virtual testing environments, assessing the viability of sustainable materials, such as graphene-based nano-material membranes for desalination, and metal–organic frameworks for harvesting water in arid regions. Optimization algorithms play a crucial role in enhancing the reliability, longevity, and cost-effectiveness of treatment and distribution facilities during design, construction, and upgrades. Additionally, AI-powered digital twins of cities enable the rapid scaling of water-sensitive urban design, integrating various systems like bioretention, collecting rainwater and storing it underground for later use, known as aquifer storage and recovery.

The synergy of AI, Internet of Things (IoT) devices, and robotics greatly enhances operational efficiency in water and wastewater facilities. Intelligent adjustment of water treatment processes, informed by real-time sensor data, ensures compliance with drinking water standards. Similarly, wastewater treatment benefits from self-adaptive unit processes, improving efficiency through real-time optimization of organic and inorganic content. Advanced technologies like anaerobic digesters and classification and sorting systems enhance biogas production and biosolid efficacy for agricultural reuse. Smart distribution systems, utilizing machine learning models with real-time data, optimize flow pressure and velocity, improving energy efficiency through autonomous control of water pump stations. These systems optimize storage usage and expedite alerts, thereby mitigating sewage overflows during wet weather events.

Utilizing predictive analytics empowered by sensor technology and cloud computing enables the detection of anomalies, pinpointing of leaks, and prioritization of repairs according to severity for immediate action. Machine learning models, in collaboration with conventional CCTV data, forecast, diagnose, and resolve defects and blockages within wastewater networks. Artificial intelligence prolongs the lifespan of assets and enhances capital expenditure optimization by automating maintenance tasks and formulating predictive upgrade plans derived from both historical and real-time evaluations.

2.4.3 Improvements in Meeting Water Demand at the User Level

At the community level, computational intelligence offers a promising avenue for enhancing sustainability, resilience, and fair water distribution. By leveraging AI-driven analysis of historical data, smart meter readings, satellite imagery, and forecasts of water consumption, precise allocations can be determined to manage competing demands across sectors and borders while ensuring compliance with withdrawal regulations. In the agricultural sector, AI applications optimize irrigation volumes and schedules to achieve optimal crop yields under varying conditions. This is further facilitated by digital twin and robotic technologies for precision farming. Autonomous analysis of satellite or drone hyperspectral imaging, supported by computer vision and machine learning algorithms, produces detailed maps of soil moisture and crop conditions. This empowers water authorities to effectively monitor and adjust supply allocations as needed.

At the household level, implementing intelligent water-saving devices such as smart toilets, taps, and sprinklers has proven to significantly decrease household water consumption. The integration of smart meters, along with predictive demand and pricing analytics, encourages behavioral shifts toward water conservation. Additionally, artificial intelligence manages decentralized potable water, stormwater, and sewerage systems, which include automated rainwater tanks, domestic water recycling systems, and home biodigesters. By incorporating real-time fluorescence sensors, combined with machine learning in household units, it becomes possible to accurately forecast and address instances of fecal contamination in drinking water. This approach aligns with World Health Organization risk standards, effectively mitigating the risk of disease outbreaks prevalent in both affluent and low-income regions.

While these solutions often rely on existing water infrastructure, AI serves as a critical catalyst in advancing water equity. Neural environmental systems models (ESMs) and optimization algorithms support international development organizations and governments in strategically prioritizing efforts related to water, sanitation, and hygiene (WASH). This strategic approach addresses urgent challenges and strengthens climate resilience within communities.

Innovative water technologies, such as solar-powered 'water ATMs' and portable devices like 'smart handpumps,' offer promising solutions for enhancing access to safe water. These technologies can be distributed and monitored remotely, potentially benefiting women and girls, who are often disproportionately affected by water scarcity. Leveraging the prevalence of personal smartphones in developing countries, these technologies enable widespread communication on issues like drinking water safety and menstrual hygiene practices. Additionally, the deployment of portable AI systems, trained to evaluate drinking water quality based on factors like free residual chlorine content, plays a crucial role in preventing waterborne diseases in urban areas.

2.5   Case Studies and Real-World Implementations

In the search for sustainable water conservation methods, the incorporation of AI has risen as a revolutionary force, transforming the landscape of water resource management and preservation. As we explore AI in water conservation, it becomes imperative to explore tangible proof of its efficacy through case studies and real-world implementations. These practical instances offer valuable insights into the real-world applications of AI, showcasing its capacity to transform water management globally. This section sets the stage to investigate into diverse case studies and implementations, offering a comprehensive exploration of how AI is making waves in the field of water conservation, addressing challenges, and facilitating the path toward a sustainable and water-secure future. Table 2.2 presents real-life examples based on applications of AI.

In summary, the collaboration between AI and water conservation holds immense promise in addressing the growing challenges linked to the management of our precious water resources. The far-reaching influence of AI, spanning intelligent water systems to data-driven insights, is fundamentally altering the course of water conservation methodologies. It is imperative to embrace these technological breakthroughs, adopting a cautious approach to ensure that AI becomes a positive catalyst in our collective efforts to safeguard and responsibly manage our essential water reservoirs.

Table 2.2 Real-life examples offer valuable insights into the practical applications of AI

Problems	AI Algorithms and their websites	Uses
An AI-Powered Leak Detection and Water Conservation System Debuts in the United States	https://wint.ai/	The WINT Water Intelligence solution serves dual objectives: firstly, to mitigate water leak damages and secondly, to enhance sustainability by substantially curbing water usage. Utilizing cutting-edge artificial intelligence technology, WINT conducts instantaneous analysis of water flow, detecting irregularities, wastage, and leaks in real-time. Upon detection, it promptly issues alerts and initiates automatic water shut-off to avert potential damages. Additionally, WINT provides comprehensive analytics and reports, empowering sustainability and facility management teams to efficiently manage consumption and mitigate risks
A Comparative Analysis of Hybrid-Wavelet Artificial Intelligence Models for Forecasting Monthly Groundwater Depth in Extremely Arid Regions of Northwest China	The research illustrates the viability of employing a wavelet-artificial neural network (WA-ANN) and wavelet-support vector regression (WA-SVR) for this forecasting task (Yu et al., 2018)	In this study focused on managing water resources, explores into the crucial endeavor of forecasting groundwater depth (GWD) for periods of 1, 2, and 3 months within the Ejina Basin. The research findings suggest that WA-SVR could prove to be a valuable tool for predicting groundwater levels, especially in situations involving ecological water conveyance
Provides appropriate treatment measures	https://www.emagin.ca/	EMAGIN utilizes AI to provide precise and timely data, resulting in cost and water savings. The software offers insights into water pollutants and suggests appropriate treatment measures. Additionally, it enhances the efficiency of wastewater cleaning and system management to prevent overflows

(continued)

Table 2.2 (continued)

Problems	AI Algorithms and their websites	Uses
Detect dangerous particles and bacteria in water	Convolution Neural Network (CNN) through Caffe Framework https://cleanwaterai.com/#	Clean Water AI functions in real-time without interruption, employing a digital microscope connected to a laptop operating on the Ubuntu system. It effectively assesses water quality and identifies contaminated areas on a map. In regions confronting ongoing clean water difficulties, this simple testing system can greatly contribute to disease prevention, potentially preserving numerous lives
Digital Risk Protection and real-time analysis of public data sources	https://www.alto-analytics.com/en_US/googles-ai-image-recognition-does-not-work-but-for-good-reason	Alto Analytics utilized AI-driven image recognition analysis of toilets to accurately measure the global impact of inadequate sanitation conditions. By utilizing open data from Dollar Street, which includes images categorizing living conditions according to income levels, their data science team uncovered a correlation between Google's AI accuracy and family income. The results suggest that 30.3% of toilets worldwide cannot be identified by AI
Smart water management	https://www.sew.ai/index.html	Smart Energy Water (SEW), an organization specializing in smart water management, utilizes AI-powered products such as Smart Customer Mobile and SmartiQ to enhance water-use efficiency. By utilizing Smart Customer Mobile, users have the opportunity to reduce water usage, engage in conservation initiatives, and obtain refunds. Additionally, the platform simplifies the process of reporting water wastage and leaks through any mobile device

(continued)

Table 2.2 (continued)

Problems	AI Algorithms and their websites	Uses
Predict ruptures, pinpoint leaks, and cut Non-Revenue Water in just 45 days	ML based algorithms https://www.fracta.ai/	Fracta, a startup based in California, employs artificial intelligence to identify leaks in pipelines and valves, aiding in improved asset management and maintenance decision-making
Management of buried pipe infrastructure by creating micro-robots specifically designed for underground pipe networks and hazardous locations	ML based algorithms https://pipebots.ac.uk/	Pipebots, a project based in the UK, endeavors to create AI-powered micro-robots for the examination and cost-effective repair of underground pipelines
Sewer and drain installation and repair	https://sewerspy.com/	A worldwide technology enterprise IBM, has employed AI and machine learning to create a prototype for in-pipe inspection named Sewer Spy, designed to detect pipeline erosion and corrosion

Funding This study was funded by the Department of Science and Technology, Government of India [DST/PRC/CPR/IITIndore (G)] for the project entitled "Technological Innovation and Intellectual Property"

References

https://www.allerin.com/blog/4-ways-ai-is-helping-with-water-management

https://www.prnewswire.com/news-releases/artificial-intelligence-powered-leak-detection-and-water-conservation-system-launches-in-the-us-300806874.html

Mahardhika, S. P., & Putriani, O. (2023, June). Deployment and use of Artificial Intelligence (AI) in water resources and water management. In *IOP Conference Series: Earth and Environmental Science* (Vol. 1195, No. 1, p. 012056). IOP Publishing.

World Food and Agriculture – Statistical Yearbook 2021 |Policy Support and Governance| Food and Agriculture Organization of the United Nations (fao.org)

Yu, H., Wen, X., Feng, Q. et al. Comparative Study of Hybrid-Wavelet Artificial Intelligence Models for Monthly Groundwater Depth Forecasting in Extreme Arid Regions, Northwest China. Water Resour Manage 32, 301–323 (2018).

Chapter 3
AI for Water Treatment

Abstract The chapter explores the integration of AI in water treatment, emphasizing its transformative impact on addressing challenges within water treatment processes. The abstract provides a concise overview of AI's pivotal role, exploring how AI algorithms and data analytics enhance the efficiency and precision of water treatment. Real-world applications and case studies are presented, illustrating instances where AI has revolutionized traditional methodologies. Emphasizing the tangible benefits, the abstract highlights AI's capacity to predict water quality, optimize treatment parameters, and autonomously adapt to dynamic conditions. As readers engage with this exploration, they gain insights into the practical implementation of AI, witnessing its potential to significantly enhance the accessibility and quality of clean water. The abstract concludes by highlighting the crucial role of AI in advancing water treatment methodologies, setting the stage for an in-depth examination.

3.1 AI in Water Treatment Applications

The integration of Artificial Intelligence into various domains has revolutionized traditional practices and introduced innovative solutions. One such transformative area is water treatment, where AI applications hold the potential to address complex challenges in water management. This introduction provides a summary of AI's role in water treatment, underscoring its importance in streamlining operations, ensuring water quality, and contributing to sustainable water resource management. As technological advancements continue to shape the future, exploring the varied applications of AI in the field of water treatment becomes crucial for fostering efficiency, resilience, and environmental conservation.

© The Author(s), under exclusive license to Springer Nature Switzerland AG 2024   31
M. K. Goyal et al., *AI Innovation for Water Policy and Sustainability*,
SpringerBriefs in Water Science and Technology,
https://doi.org/10.1007/978-3-031-72014-7_3

3.2   AI-Enhanced Water Quality Monitoring

In the modern era, inadequate water quality remains a major concern, with approximately one in nine people worldwide relying on unsafe sources for drinking water
(WHO, UNICEF, 2015). Additionally, approximately 90% of untreated sewage in
developing nations is released into waterways (UNESCO WWAP, 2015). Addressing
these water quality challenges necessitates the adoption of innovative technologies.
Table 3.1 illustrates various scenarios for the utilization of AI.

The integration of AI into water quality monitoring has marked a transformative
leap forward in our ability to comprehend, predict, and manage the complex dynamic
forces of water ecosystems. The applications of AI, ranging from predictive analytics
and anomaly detection to optimizing treatment processes, have demonstrated their
efficacy in providing timely and accurate insights. AI's adaptability, particularly
in leveraging advanced models like Recurrent Neural Network (RNN) and Long
Short-Term Memory network (LSTM), has proven invaluable in forecasting pollutant concentrations and system performance. Furthermore, AI's role in anomaly
detection, identifying irregularities such as equipment malfunctions, showcases its
potential for proactive maintenance and operational resilience. As we navigate the
challenges of water quality management, the continued integration and refinement
of AI technologies hold the promise of more sustainable and efficient water resource
utilization, ultimately contributing to the safeguarding of our vital water ecosystems.
The journey towards AI-enhanced water quality monitoring is not just a technological
advancement but a crucial step toward ensuring the long-term health and resilience
of our precious water resources.

In the domain of water treatment, the utilization of AI and ML has presented
significant effectiveness in managing complex relationships and providing accurate
predictions, especially when dealing with data showing intricate nonlinear patterns
that are difficult for traditional mathematical models to handle. Research conducted
by Li et al. (2021) highlights the successful implementation of AI across various
water treatment domains. These include predicting water quality, estimating coag-
ulant doses, controlling membrane fouling, identifying potential disinfection by-
products, and optimizing membrane preparation processes. Integrating AI technologies provides new insights, improving comprehension of intricate water treatment
processes and providing essential decision-making assistance for water treatment
management systems.

However, despite the numerous advantages, the broad implementation of AI technologies in practical water treatment faces several obstacles. As noted by Alam
et al. (2022), critical issues include ensuring data availability and making appropriate selections, addressing concerns regarding poor reproducibility, and the necessity for conclusive evidence of successful applications in real-world water treatment scenarios. Overcoming these challenges is crucial to fully harnessing the
transformative potential of AI in revolutionizing water treatment processes.

Table 3.1 Utilization of AI in different challenging scenarios

Sl. No	Utilizations	Outcome
1	Clean Water AI	Clean Water AI, an innovative American system, utilizes AI to address SDGs related to water quality. It integrates IoT and Convolutional Neural Networks (CNN) for instant analysis, identifying contaminants, including bacteria, even without an internet connection. This system employs cost-effective, readily available components and is currently accessible for $500, with an anticipated price reduction as AI technology advances and gains global acceptance
2	Forecasting water quality in Iran (Talesh et al., 2019)	In the Sefidrud basin of Iran, the Water Quality Index (WQI) is calculated using Support Vector Machines (SVM) models, which are established through laboratory analysis of water samples. The SVM models effectively analyze 87% of the variability in the overall water quality index. Beyond establishing the WQI, the SVM results can enhance river management practices for improved water quality
3	Monitoring water quality in Africa using satellite technology	SERVIR has utilized deep learning techniques and satellite data to examine past alterations in water quality within inland and trans-boundary lakes across various African nations, such as Kenya, Malawi, Rwanda, Tanzania, and Uganda. Through a web-based decision information system, users can access data on parameters like chlorophyll-a levels, lake surface temperature, and suspended matter. This system supports efforts aimed at addressing pollutants that compromise water quality
4	Disinfection techniques employing AI/ML algorithms	Research has demonstrated the effectiveness of AI methods in controlling chlorination, while machine learning models have proven successful in simulating concentrations of disinfection byproducts (DBPs). While pre-chlorination peroxide/ozonation has been explored in certain studies, chlorine remains a commonly used disinfectant in the treatment of surface waters across drinking water facilities. Researchers have also noted its efficacy in modeling DBP concentrations in consumer taps, residences, and treated water distribution networks (Godo-Pla et al., 2021; Peleato et al., 2022; Babaei et al., 2023; Hu et al., 2023)
5	AI/ML methods applied in adsorption	Adsorption methods are commonly used in water and wastewater treatment to eliminate pollutants and impurities. This process involves transferring target molecules, known as adsorbates, from a fluid to a solid surface, known as an adsorbent or sorptive medium. To further improve the process, researchers have developed models that utilize machine learning to produce substantial forecasts during the adsorption phase. Employing machine learning for adsorption processes could assist operators in making informed decisions (Bhagat et al., 2021; Zhao et al., 2022)

(continued)

Table 3.1 (continued)

Sl. No	Utilizations	Outcome
6	AI and ML methodologies applied in membrane filtration	In the field of water and wastewater treatment, membrane processes involve the filtration or separation of impurities using barriers. Machine learning has been applied in various ways to predict, simulate, and improve membrane filtration processes. Scientists have created models for reverse osmosis, ultrafiltration, nanofiltration, and microfiltration (Waqas et al., 2022; Li et al., 2022)
7	AI and ML methodologies employed in the management of water quality	Research has explored machine learning and artificial intelligence methods in the field of water quality management. Studies have demonstrated the utility of ML models in forecasting and simulating parameters relevant to water quality management (Mathaba & Banza 2023); Aslam et al., 2022; Ziyad Sami et al., 2022; Pham et al., 2021)

3.3   Predictive Modeling for Water Treatment Optimization

In the domain of water treatment, the need for effective and streamlined processes has grown significantly, driven by the escalating challenges in water resource management. Predictive modeling, a robust analytical approach, stands out as a pivotal method set to transform the landscape of water treatment optimization. Using sophisticated algorithms and insights derived from data analysis, predictive modeling can elevate decision-making processes within water treatment facilities. This introduction delves into the sphere of Predictive Modeling for Water Treatment Optimization, examining its importance, applications, and the potential transformative influence it holds in ensuring the sustainable and efficient stewardship of this indispensable resource. Addressing the challenge of anomaly detection in large datasets, especially in the domain of water treatment, has prompted the exploration of advanced machine learning techniques across multiple studies. The significant challenge of unbalanced data, Characterized by infrequent irregularities in dynamic sensor data, a water quality anomaly detection challenge was conducted during the Genetic and Evolutionary Computation Conference (GECCO).

Various approaches were presented in different studies, incorporating Convolutional Neural Networks (CNN), LSTMs, and deep networks, showcasing their effectiveness in water quality anomaly detection. The comparison across these studies proves challenging due to variations in datasets, models, and parameters employed. However, deep learning methods demonstrated superior accuracy in feature learning and reduced false positive rates compared to traditional machine learning methods, marking a significant stride in anomaly detection capabilities.

Detecting anomalies in water quality is challenging due to the diverse range of anomaly types, including sudden spikes, low variability, constant offset, Abrupt change, fluctuations in data, gradual deviation, and minor rapid increases (Fu et al.,

2022). Investigations utilizing Long Short-Term Memory (LSTM) network for classifying different categories of anomalies in river water quality have demonstrated varying detection abilities among parameters and monitoring locations. LSTM models have been shown to reduce false detection rates compared to regression-based ARIMA approaches. Detection of water quality anomalies has traditionally relied on analyzing multivariate time series data collected from water quality sensors, distinguishing it from the detection of leaks, which typically involves different types of data. Moreover, machine learning models have been explored to predict water quality parameter values by analyzing disparities between predicted and observed data for anomaly detection. Recent advancements include the utilization of Generative Adversarial Networks (GAN) and Graph Neural Networks (GNN). GAN models have shown promise in addressing complex anomaly detection challenges, while attention-based GNNs offer interpretability for identified anomalies, highlighting the significance of neighboring nodes (sensors) in understanding node behavior (Fu et al., 2022).

In summary, the adoption of deep learning techniques such as LSTM, GAN, and GNN for detecting contamination in water treatment has greatly improved with the availability of high-quality open data. Nevertheless, further validation through testing on real-world multivariate data is crucial to refine and validate the effectiveness of these methods.

3.4 Intelligent Water Treatment Systems and Technologies

In the twenty-first century, poor water quality remains a significant challenge, with approximately one in nine people worldwide relying on unsafe and unimproved water sources for drinking (WHO, UNICEF, 2015). Furthermore, an alarming 90% of wastewater in developing nations is released untreated into waterways (UNESCO WWAP, 2015). Addressing these water quality challenges requires the adoption of disruptive technologies.

Across different domains, the integration of AI has led to notable progress in water quality monitoring. Internationally, AI is being utilized to address various spatial and temporal challenges related to water quality. These applications encompass a wide spectrum of functionalities, which can be classified into the following categories:

- Continuous water quality monitoring using AI and IoT: AI-powered IoT solutions facilitate frequent data collection and predictive networks for real-time monitoring of water quality. Usually deployed upstream of water bodies, these systems monitor and forecast parameters like dissolved oxygen and total organic carbon over different time intervals, ranging from short-term (hourly) to seasonal (during wet and dry seasons).
- Evaluating water quality via sampling: AI innovations in pattern recognition and sensor tech accelerate the identification of bacterial pollutants in water samples. These AI-driven tools offer a streamlined alternative to traditional

manual methods, improving the interpretation of color-based indicators and enabling the pinpointing of specific contaminants, diseases, and infections.

- Utilizing AI and remote sensing (satellite imagery) for monitoring the quality of extensive water bodies: AI is employed to analyze satellite images to categorize water bodies in remote regions where traditional sensor installation is unfeasible. This approach assists in detecting changes or trends in water quality indicators over time. Typical parameters monitored include total suspended solids, chlorophyll-a, diffuse attenuation coefficient, sea surface temperature, and fluorescence line-height.

Table 3.2 presents the various predicting methodologies used in water treatments implying AI application (Ismail et al., 2021). It emphasizes the role of AI applications, notably the use of artificial neural networks (ANNs). ANNs are commonly utilized for predicting contaminant removal across different treatment processes, thanks to their self-adaptive and self-learning abilities. With access to suitable training algorithms and ample data, ANNs can effectively tackle complex multivariate nonlinear problems. This makes them valuable tools in experimental designs aimed at pollutant removal in both water and wastewater treatment.

Moreover, deep learning has become a pivotal technology in wastewater treatment, introducing innovative solutions to enhance various facets of the process (Mehmood

Table 3.2 Different methodologies are employed in water treatment

Artificial Neural Network (ANN)
Multi-layer perception (MLP) Feed forward neural network (FFNN) Fuzzy neural network (FNN) Linguistic Models (LM)
Fuzzy-based
Fuzzy logic (FL) Adaptive neuro-fuzzy interference system (ANFIS) Dynamic evolving neural-fuzzy inference system (DENFIS) Fuzzy inference systems (FIS)
Deep learning (DL)
Stacked denoising auto-encoders (SDAE) deep learning network
Bayesian Belief Networks (BBNs)
Least squares support vector machine (LSVM)
Multiple Rough Classifier Systems (MRCS)
Support Vector Machine (SVM)
K-nearest neighbour (k-NN)
Grey dynamic modelling and genetic algorithm
Reinforcement learning technique
Convolutional neural network (CNN)
Long short-term memory networks (LSTM)

et al., 2020). A noteworthy application lies in predictive analytics, where deep learning architectures like Recurrent Neural Network (RNN) or Long Short-Term Memory network (LSTM), prove instrumental in forecasting pollutant concentrations, water quality parameters, or system performance. These models excel at capturing intricate temporal dependencies within wastewater treatment data, enabling more accurate and timely predictions. Additionally, deep learning plays a crucial role in anomaly detection, identifying irregularities like equipment malfunctions or unexpected variations in pollutant levels. This aids in proactive maintenance, minimizing operational disruptions. The capacity of deep learning algorithms to analyze and learn from extensive datasets contributes to optimizing treatment processes, enhancing efficiency, and ultimately advancing the effectiveness of wastewater treatment systems.

In summary, the advent of Intelligent Water Treatment Systems and Technologies marks a transformative era in water management. The integration of artificial intelligence and advanced technologies has paved the way for more efficient, adaptive, and sustainable approaches to addressing the complexities of water treatment. From predictive analytics and anomaly detection to optimizing treatment processes, these intelligent systems showcase their prowess in enhancing the precision and effectiveness of water treatment operations. As we move forward, continued research, innovation, and implementation of intelligent water treatment solutions will play a pivotal role in ensuring the availability of clean and sustainable water resources for our growing global population. Embracing these technologies is not just a necessity but a strategic imperative for achieving resilient and smart water management practices in the face of evolving environmental challenges.

3.5 Case Studies of AI Success in Water Treatment

In the modern era, inadequate water quality poses a substantial challenge, as evidenced by the fact that one in nine people worldwide relies on unsafe drinking water sources (WHO, UNICEF, 2015). Furthermore, approximately 90% of untreated sewage is released into water bodies in developing nations (UNESCO WWAP, 2015). Addressing these water quality challenges necessitates the adoption of innovative technologies. Table 3.3 presents a list of successful case studies illustrating the application of AI in water treatment.

In conclusion, the showcased case studies on the triumphs of AI in water treatment underscore the transformative capacity of AI to revolutionize water management practices. From predictive analytics enhancing water quality monitoring to machine learning-driven optimization of treatment processes, these applications signify a profound shift in how we approach challenges in water treatment.

These success stories illuminate the adaptability of AI technologies across diverse contexts, demonstrating their efficacy in addressing critical issues like contamination detection, resource optimization, and infrastructure resilience. As we traverse through the complex environment of water treatment, AI emerges as a potent ally,

Table 3.3 Case Studies of AI Implementation in Water Treatment

Sl. No	Utilizations	Outcome
1	Clean Water AI	Clean Water AI, an innovative American system, utilizes AI to address SDGs related to water quality. It integrates IoT and Convolutional Neural Networks (CNN) for instant analysis, identifying contaminants, including bacteria, even without an internet connection. This system employs cost-effective, readily available components and is currently accessible for $500, with an anticipated price reduction as AI technology advances and gains global acceptance
2	Forecasting water quality in Iran (Talesh et al., 2019)	In the Sefidrud basin of Iran, the Water Quality Index (WQI) is determined using Support Vector Machines (SVM) models. These models are developed based on laboratory analysis of water samples. The SVM models effectively analyze 87% of the variability in the overall water quality index. Beyond establishing the WQI, the SVM results can enhance river management practices for improved water quality
3	Monitoring water quality in Africa using satellite technology	SERVIR uses deep learning and satellite data to track water quality changes in African lakes like those in Kenya, Malawi, Rwanda, Tanzania, and Uganda. Their web-based system provides data on chlorophyll-a, lake temperature, and suspended matter, aiding pollution mitigation efforts
4	Prediction of groundwater quality in India (Kadam et al., 2019)	Artificial Neural Network and Multiple Linear Regression models were employed to assess groundwater potability within the Shivganga River basin using a WQI, which incorporated diverse physicochemical parameters such as pH, EC, TDS, Ca, Mg, Na, K, Cl, HCO, TH, SO, PO and NO. The efficacy of this model was demonstrated during both pre- and post-monsoon periods, indicating its suitability for application in comparable regional settings for groundwater quality monitoring
5	Improving access to safe drinking water in Ghana using satellite technology	The Water Resources Commission of Ghana utilizes the Africa Regional Data Cube (ARDC) to bolster water quality assessment in the Weija Reservoir, a critical freshwater source for Accra and its environs. Within the ARDC framework, various algorithms are deployed, including the NASA Chlorophyll-A detection algorithm.
6	AI-driven sewer monitoring in the United States	SmartCover Systems, an American company, employs AI and IoT technology for monitoring water and wastewater infrastructure. Using satellite communication, the system gathers data, detects blockages and stormwater infiltration, and offers real-time protection insights through AI-powered trend analysis
7	Assessing the state and potential hazards of the drinking water distribution system in the United States	Fracta, an American company, improves water infrastructure management with AI-powered systems. Their solutions evaluate the condition and risk of drinking water distribution mains using machine learning algorithms to calculate the Likelihood of Failure (LOF). This innovative approach transforms asset management from reactive to proactive, reducing water main breaks, minimizing non-revenue water (NRW), optimizing leak detection, valve maintenance efforts, and providing stakeholders with valuable insights into infrastructure risks and costs

providing innovative solutions that not only enhance efficiency but also significantly contribute to the sustainability of water resources.

Looking ahead, these case studies serve as inspiring guides for further research, stage, and execution of AI powered solutions in the realm of water treatment. The collaboration between technology and water management holds great promise, underscoring the importance of harnessing the potential of AI to meet the escalating demands for clean and accessible water worldwide. With sustained exploration and integration, AI is positioned to play a pivotal role in shaping a future where water treatment is not merely a necessity but a triumph of innovation and intelligent problem-solving.

Funding This study was funded by the Department of Science and Technology, Government of India [DST/PRC/CPR/IITIndore (G)] for the project entitled "Technological Innovation and Intellectual Property"

References

Alam, G., Ihsanullah, I., Naushad, M., & Sillanpää, M. (2022). Applications of artificial intelligence in water treatment for optimization and automation of adsorption processes: Recent advances and prospects. *Chemical Engineering Journal, 427,* 130011.

Aslam, B., Maqsoom, A., Cheema, A. H., Ullah, F., Alharbi, A., & Imran, M. (2022). Water quality management using hybrid machine learning and data mining algorithms: An indexing approach. IEEE Access, 10, 119692–119705.

Babaei, A. A., Tahmasebi Birgani, Y., Baboli, Z., Maleki, H., & Ahmadi Angali, K. (2023). Using water quality parameters to prediction of the ion-based trihalomethane by an artificial neural network model. Environmental Monitoring and Assessment, 195(8), 917.

Bhagat, S. K., Pyrgaki, K., Salih, S. Q., Tiyasha, T., Beyaztas, U., Shahid, S., & Yaseen, Z. M. (2021). Prediction of copper ions adsorption by attapulgite adsorbent using tuned-artificial intelligence model. Chemosphere, *276,* 130162.

Fu, G., Jin, Y., Sun, S., Yuan, Z., & Butler, D. (2022). The role of deep learning in urban water management: A critical review. *Water Research*, 118973.

Godo-Pla, L., Rodríguez, J. J., Suquet, J., Emiliano, P., Valero, F., Poch, M., & Monclús, H. (2021). Control of primary disinfection in a drinking water treatment plant based on a fuzzy inference system. Process Safety and Environmental Protection, *145,* 63–70.

https://www.unesco.org/en/wwap

https://www.unicef.org/reports/unicef-annual-report-2015

Hu, G., Mian, H. R., Mohammadiun, S., Rodriguez, M. J., Hewage, K., & Sadiq, R. (2023). Appraisal of machine learning techniques for predicting emerging disinfection byproducts in small water distribution networks. *Journal of Hazardous Materials, 446,* 130633.

Ismail, W., Niknejad, N., Bahari, M., Hendradi, R., Zaizi, N. J. M., & Zulkifli, M. Z. (2021). Water treatment and artificial intelligence techniques: a systematic literature review research. *Environmental Science and Pollution Research*, 1–19.

Kamyab-Talesh, F., Mousavi, S. F., Khaledian, M., Yousefi-Falakdehi, O., & Norouzi-Masir, M. (2019). Prediction of water quality index by support vector machine: a case study in the Sefidrud Basin, Northern Iran. Water Resources, 46, 112-116.

Li, B., Yue, R., Shen, L., Chen, C., Li, R., Xu, Y., ... & Lin, H. (2022). A novel method integrating response surface method with artificial neural network to optimize membrane fabrication for wastewater treatment. *Journal of Cleaner Production, 376,* 134236.

Li, L., Rong, S., Wang, R., & Yu, S. (2021). Recent advances in artificial intelligence and machine learning for nonlinear relationship analysis and process control in drinking water treatment: A review. *Chemical Engineering Journal, 405,* 126673.

Mathaba, M., & Banza, J. (2023). A comprehensive review on artificial intelligence in water treatment for optimization. Clean water now and the future. Journal of Environmental Science and Health, Part A, 58(14), 1047–1060.

Mehmood, H., Mukkavilli, S. K., Weber, I., Koshio, A., Meechaiya, C., Piman, T., ... & Liao, D. (2020). Strategic Foresight to Applications of Artificial Intelligence to Achieve Water-related Sustainable Development Goals. *United Nations University Institute for Water, Environment and Health, Hamilton, Canada. UNU-INWEH Report Series.*

Peleato, N. M. (2022). Application of convolutional neural networks for prediction of disinfection by-products. Scientific Reports, 12(1), 612.

Pham, Q. B., Mohammadpour, R., Linh, N. T. T., Mohajane, M., Pourjasem, A., Sammen, S. S., ... & Nam, V. T. (2021). Application of soft computing to predict water quality in wetland. Environmental Science and Pollution Research, 28, 185–200.

UNESCO WWAP. (2015). *The United Nations World Water Development Report 2015: Water for a Sustainable World.* Paris: UNESCO World Water Assessment Programme.

Waqas, S., Harun, N. Y., Sambudi, N. S., Arshad, U., Nordin, N. A. H. M., Bilad, M. R., ... & Malik, A. A. (2022). SVM and ANN modelling approach for the optimization of membrane permeability of a membrane rotating biological contactor for wastewater treatment. Membranes, 12(9), 821.

WHO, UNICEF. (2015). Progress on Sanitation and Drinking Water: 2015 Update and MDG Assessment. World Health Organization and United Nations Children's Fund Joint Monitoring Programme for Water Supply and Sanitation (JMP). Geneva: World Health Organization.

Zhao, Y., Fan, D., Li, Y., & Yang, F. (2022). Application of machine learning in predicting the adsorption capacity of organic compounds onto biochar and resin. Environmental Research, 208, 112694.

Ziyad Sami, B. F., Latif, S. D., Ahmed, A. N., Chow, M. F., Murti, M. A., Suhendi, A., ... & El-Shafie, A. (2022). Machine learning algorithm as a sustainable tool for dissolved oxygen prediction: a case study of Feitsui Reservoir, Taiwan. Scientific Reports, 12(1), 3649.

Chapter 4
AI for Water Policy

Abstract The chapter explores the transformative function of AI in the field of water policy. As water-related challenges escalate globally, the abstract provides a comprehensive overview of how AI technologies contribute to shaping and optimizing policies for effective water management, conservation, and sustainability. Key areas of focus include data-driven decision-making, predictive modeling, and adaptive governance, illustrating how AI enhances the precision and responsiveness of water policies. Real-world applications and case studies are explored, highlighting instances where AI has revolutionized traditional policy approaches. Emphasizing the tangible impact, the abstract underlines AI's potential to address emerging water issues, adoptive resilience, and facilitate dynamic policy strategies. Policymakers, researchers, and practitioners engaging with this chapter gain valuable insights into the practical implementation of AI, offering a roadmap for leveraging its capabilities to navigate and address the complex challenges inherent in water policy formulation and implementation.

4.1 Introduction to AI in Water Policy

Explore the untapped promise of artificial intelligence in water policy, where cutting-edge technologies and responsible governance collide. Policymakers are better equipped to make educated decisions that strike a careful balance between preserving the environment and meeting human needs when they utilise data-driven insights. AI systems simplify complicated frameworks by deftly navigating the difficult terrain of water policy. These technologies foresee patterns in water usage through predictive analytics, giving authorities the means to take preemptive measures to address issues like pollution and scarcity. By utilising artificial intelligence algorithms and comprehensive data analysis methodologies, water utilities may maximise the utilisation of information and data. Better decision-making, enhanced service delivery, and cost savings are the results of this optimisation.

The integration of AI into the domain of water policy represents a transformative paradigm, introducing a fresh era characterized by decision-making driven by data

M. K. Goyal et al., *AI Innovation for Water Policy and Sustainability*,
SpringerBriefs in Water Science and Technology,
https://doi.org/10.1007/978-3-031-72014-7_4

and strategic governance. As the global challenges related to water scarcity, quality management, and sustainable resource utilization intensify, the application of AI technologies emerges as a promising avenue to address these complexities. In the realm of water policy, AI offers unprecedented opportunities to enhance efficiency, optimize resource allocation, and devise innovative strategies for sustainable water governance. This investigation into "AI in Water Policy" explores the convergence of artificial intelligence and policy development in the water domain. The relevance of AI in water policy spans a spectrum of crucial aspects, including but not limited to regulatory frameworks, resource allocation, disaster preparedness, and equitable distribution. Through advanced algorithms, predictive analytics, and machine learning models, AI empowers policymakers to analyze vast datasets, predict trends, and formulate evidence-based policies that foster resilience and sustainability in water management.

This discussion aims to unravel the intricacies of how AI technologies are influencing the landscape of water policy. From optimizing water allocation to mitigating the influence of climate change on water resources, the integration of AI presents a myriad of possibilities to revolutionize the way societies approach water governance. By fostering a deeper understanding of AI's role in shaping water policy, this exploration seeks to contribute valuable insights to policymakers, researchers, and stakeholders navigating the complex terrain of water resource management in the twenty-first century.

AI is crucial in the optimization of resource allocation, particularly in the domain of water distribution. Through the implementation of smart algorithms, AI ensures the precise allocation of water resources, minimizing waste and promoting equitable access for every community. The real-time monitoring capabilities of AI act as a vigilant guardian for water quality, with automated sensors swiftly detecting anomalies and enabling rapid responses to potential threats, thereby safeguarding ecosystems and public health. AI emerges as a stalwart guardian against water pollution, with smart monitoring systems swiftly identifying pollutants and allowing for targeted interventions, contributing significantly to environmental conservation. Furthermore, AI fosters biodiversity preservation by analyzing ecological data, aiding in the identification and protection of vulnerable ecosystems. The adaptability of AI-driven policy frameworks is explored, as these dynamic systems evolve based on real-time data, ensuring policies remain relevant and effective in the ever-changing landscape of water management. AI-driven policies also foster community engagement by integrating local knowledge and preferences, resonating with diverse community needs for widespread acceptance. Additionally, AI contributes to climate change resilience through advanced modeling, assisting in crafting robust strategies that anticipate and mitigate the effects of climate change on water resources. Lastly, the integration of AI promotes sustainable water practices, encouraging conservation behaviors and optimizing irrigation techniques to steer humanity towards a more sustainable and resilient future. The following section illustrates various methods through which AI can make contributions to a sustainable future.

4.1.1 AI's Role in Resource Allocation

Precision in Distribution

Witness AI's prowess in optimizing water distribution. Through smart algorithms, water resources are allocated with unparalleled precision, ensuring every community receives its fair share. This not only minimizes waste but also promotes equitable access.

Real-Time Monitoring

Experience the real-time monitoring capabilities of AI, acting as a vigilant guardian of water quality. Automated sensors detect anomalies, enabling swift responses to potential threats, safeguarding ecosystems and public health.

4.1.2 Environmental Conservation Through AI

Mitigating Pollution

Embark on a journey where AI becomes a stalwart guardian against water pollution. Smart monitoring systems identify pollutants swiftly, allowing for targeted interventions. This proactive approach contributes significantly to environmental conservation.

Biodiversity Preservation

Explore how AI fosters biodiversity preservation. By analyzing ecological data, AI aids in identifying and safeguarding vulnerable ecosystems, fostering a harmonious coexistence between human activities and nature.

4.1.3 AI in Policy Implementation

Adaptive Policy Frameworks

Delve into the adaptability of AI-driven policy frameworks. These dynamic systems evolve based on real-time data, ensuring policies remain relevant and effective amidst the ever-changing landscape of water management.

Community Engagement

Experience the inclusive nature of AI-driven policies, fostering community engagement. By integrating local knowledge and preferences, these policies resonate with the diverse needs of communities, ensuring widespread acceptance and adherence.

4.1.4 AI and Climate Change Resilience

Future-Proofing Strategies

Uncover how AI contributes to climate change resilience. Through advanced modeling, AI assists in crafting robust strategies that anticipate and mitigate the effects of climate change on water reserves.

Sustainable Practices

Explore the integration of AI in promoting sustainable water practices. From encouraging water conservation behaviors to optimizing irrigation techniques, AI is instrumental in guiding humanity towards a more sustainable and resilient future.

As we embrace the era of AI in water policy, a sustainable future unfolds. From precision in resource allocation to proactive environmental conservation, the possibilities are limitless. This intersection of technology and policy not only safeguards our water resources but paves the way for a harmonious coexistence between humanity and the environment.

4.2   Data-Driven Decision-Making in Water Governance

In the swiftly evolving arena of water governance, the integration of data-driven decision-making emerges as a game-changer. This article navigates the intersection of data, governance, and water management, highlighting the profound impact data-driven approaches have on shaping policies and ensuring sustainable practices. Data-driven decision-making provides comprehensive insights, enhancing the precision of policies and addressing unique challenges in water governance. Data-driven decision-making aids in analyzing data for long-term strategies, ensuring sustainable practices in water resource management. Data-driven policies evolve based on real-time data, ensuring adaptability to the changing landscape of water governance. By incorporating diverse datasets and community feedback, data-driven decision-making ensures policies are responsive to the unique needs of various regions and demographics. Data-driven decision-making achieves a delicate balance by considering both the immediate requirements of communities and the long-term health of ecosystems.

As we embrace the era of data-driven decision-making in water governance, a new frontier unfolds. From optimized resource allocation to proactive pollution prevention, the impact is transformative. The collaboration between data utilization and governance not only guarantees effective water management but also lays the groundwork for a sustainable and robust future.

Embark on a transformative journey where data emerges as the pivotal bridge addressing information gaps in water governance. In contrast to traditional decision-making hindered by insufficient information, data-driven approaches empower policymakers with comprehensive insights, fostering more informed and effective decisions. Witness the precision enhancement in water governance policies through data analysis, tailoring strategies to regional nuances and promoting equitable water distribution. Unlock the potential of predictive analytics to forecast water usage patterns, enabling proactive measures against scarcity and efficient resource management. Experience real-time monitoring capabilities for swift responses to threats like pollution, ensuring the preservation of water quality. Delve into the adaptability of policies crafted through data-driven decision-making, evolving in response to dynamic challenges. Embrace inclusivity by incorporating diverse datasets and community feedback, ensuring policies resonate with the unique needs of various regions and demographics. Explore how data-driven approaches serve as a shield against water pollution, preserving water quality and safeguarding ecosystems. Finally, observe the delicate balance achieved between human needs and environmental preservation, as data-driven decision-making considers both immediate community requirements and the long-term health of ecosystems, ensuring environmental sustainability.

4.2.1 Introduction to Data-Driven Decision-Making in Water Governance

Bridging the Information Gap
Embark on a journey where data serves as the bridge that spans the information gap in water governance. Traditional decision-making often faced challenges due to insufficient information. However, with data-driven approaches, policymakers gain comprehensive insights, enabling more informed and effective decisions.

Enhancing Policy Precision
Witness how data-driven decision-making enhances the precision of water governance policies. By analyzing vast datasets, authorities can tailor policies to specific regional needs, addressing unique challenges and fostering more equitable water distribution.

4.2.2 The Power of Predictive Analytics

Forecasting Water Usage Patterns
Unlock the potential of predictive analytics in forecasting water usage patterns. Data-driven models analyze historical data to predict future trends, empowering decision-makers to proactively address water scarcity and manage resources efficiently.

Real-Time Monitoring for Swift Action
Explore the real-time monitoring capabilities enabled by data-driven decision-making in water governance. Automated systems equipped with sensors detect anomalies promptly, allowing for swift responses to potential threats like pollution or infrastructure issues.

4.2.3 Optimizing Resource Allocation

Precision in Water Distribution
Discover the precision achieved in water distribution through data-driven decision-making. Algorithms analyze factors such as demand, supply, and infrastructure, ensuring optimal resource allocation. This not only minimizes waste but also promotes fair and efficient water distribution.

Strategic Planning for Resource Management
Explore how data-driven decision-making contributes to strategic planning in resource management. The ability to analyze data aids in developing long-term strategies, fostering sustainable practices that go beyond immediate concerns and address future challenges.

4.2.4   Overcoming Governance Challenges

Adaptable Policies for a Changing Landscape
Dive into the adaptability of policies crafted through data-driven decision-making. These policies evolve based on real-time data, ensuring governance frameworks remain relevant and effective in the face of dynamic challenges.

Inclusive Decision-Making
Experience the inclusivity embedded in data-driven decision-making. By incorporating diverse datasets and community feedback, policies become more responsive to the unique needs of various regions and demographics.

4.2.5   Ensuring Environmental Sustainability

Proactive Pollution Prevention
Witness how data-driven approaches act as a shield against water pollution. Early detection of pollutants through constant monitoring allows for proactive measures, preserving water quality and safeguarding ecosystems.

Balancing Human Needs and Environmental Preservation
Explore the delicate balance achieved between human needs and environmental preservation. Data-driven decision-making ensures that policies consider both the immediate requirements of communities and the long-term health of ecosystems.

In summary, the paradigm shift towards data-driven decision-making in water governance marks a transformative era in our approach to managing this precious resource. The integration of robust data analytics, technological innovations, and collaborative frameworks has laid the foundation for more informed and efficient water governance practices. Embracing data-driven solutions empowers policymakers, water agencies, and stakeholders with the insights needed to address complex challenges and make proactive decisions. As we navigate the dynamic landscape of water management, the commitment to leveraging data for sustainable practices is paramount. Embarking on a path toward data-centric governance not only improves our capacity to respond to changing water conditions but also emphasizes our duty to preserve this priceless resource for present and future communities. The era of data-driven water governance beckons us to continually refine our strategies, foster innovation, and forge partnerships to ensure a resilient and equitable water future.

4.3   Forecasting Techniques for Managing Water Resources

Forecasting Techniques for Water Resource Management is an innovative and crucial approach that harnesses the power of advanced data analytics to revolutionize the management of water resources. Amidst escalating challenges like shifting climate patterns, population growth, and rising pressures on water ecosystems, the demand for proactive and effective strategies has become imperative. Predictive analytics offers a transformative solution by utilizing historical data, sophisticated algorithms, and modeling techniques to anticipate future trends and behaviours within the realm of water resources. This emerging field holds immense promise for improving decision-making, optimizing resource distribution, and promoting the sustainable utilization of water. By leveraging predictive analytics, water resource managers can gain valuable insights into potential fluctuations in water availability, quality, and usage patterns. This proactive approach empowers stakeholders to implement preemptive measures, mitigate risks, and respond efficiently to dynamic environmental conditions.

Predictive Analytics for Water Resource Management delves into the fundamental concepts, methodologies, and applications that characterize this cutting-edge discipline. By navigating the convergence of data science, hydrology, and environmental management, these remarks establish a foundation for a thorough exploration of how predictive analytics can intricately influence the trajectory of water resource management. This exploration aims to contribute significantly to enhancing resilience, efficiency, and sustainability in addressing the ever-evolving challenges within this critical field. The integration of predictive analytics in water resource management marks a pivotal advancement in our approach to sustainable and efficient water utilization. Through the sophisticated analysis of data, predictive analytics equips decision-makers with valuable insights into future water trends, enabling proactive and informed strategies for resource allocation, conservation, and crisis response. The adoption of such technologies reflects a commitment to harnessing innovation for the greater good, addressing the challenges posed by growing water demands, climate variability, and population growth. As we stride into the future, the continued refinement and implementation of predictive analytics in water management will be crucial in guiding us toward a more resilient and water-secure future.

4.4   Adaptive Governance Strategies: Harnessing AI's Potential

In the ever-evolving landscape of governance, the integration of adaptive strategies fuelled by AI stands as a catalyst for change. This article delves into the transformative potential of harnessing AI, examining how it reshapes governance approaches, fosters flexibility, and contributes to the effectiveness of adaptive strategies. Embarking on a transformative journey, we redefine adaptive governance strategies by integrating the power of AI. In this era of rapid change, embracing the capabilities of AI becomes crucial for authorities seeking to navigate complexities and establish a governance framework that dynamically adapts to evolving circumstances. The evolution of decision-making in governance is evident as intelligent algorithms drive adaptive strategies. This infusion of AI empowers authorities to make swift and informed decisions, responding to emerging challenges with unparalleled agility and precision. As we witness this revolution, the synergy between adaptive governance and AI becomes a cornerstone for effective and future-ready governance systems.

4.4.1   Enhancing Flexibility

Unlocking a new era in policy implementation, AI shows a pivotal function in enhancing the agility of adaptive governance. By processing extensive datasets in real-time, AI ensures policies remain dynamic and responsive to evolving conditions, thus fostering a flexible and adaptive governance structure. Additionally, delve into the realm of responsive resource allocation powered by AI. Intelligent algorithms meticulously analyze data patterns, enabling efficient allocation of resources that caters to immediate needs while seamlessly adapting to evolving priorities. Together, these advancements highlight the transformative impact of AI in shaping a more agile and responsive approach to governance and resource management.

4.4.2   AI's Role in Crisis Management

In the realm of crisis management, AI shows a crucial role in both proactive prevention and adaptive response within the framework of adaptive governance. Through the lens of predictive analytics, AI emerges as a proactive force by identifying potential crisis scenarios before they unfold. This foresight empowers authorities to implement preventative measures, significantly reducing the impact on governance structures. Moving into the adaptive crisis response phase, AI continues to shine. By continuously analyzing real-time data during a crisis, AI enables authorities to dynamically tailor responses. This adaptability ensures that governance strategies remain effective, even in the face of unforeseen challenges. The combination of proactive prevention and adaptive response showcases the transformative potential of AI in crisis management, enhancing the resilience and efficacy of governance systems.

4.4.3   Overcoming Governance Challenges

Navigating the intricate landscape of governance challenges is significantly enhanced through the utilization of AI. This technology, with its adaptive strategies, efficiently processes multifaceted data to provide nuanced insights for effective decision-making by authorities. The inclusive nature of AI-driven adaptive governance is a transformative aspect. By analyzing diverse datasets and incorporating community input, decision-making becomes more democratic. This inclusivity ensures that policies are aligned with the diverse needs of the population, fostering a governance approach that is responsive and representative of the broader community.

AI enriches adaptive governance by facilitating well-informed and quick decision-making, fostering flexibility in policy implementation, and contributing to the prevention and response to crises. In the realm of decision-making, AI promotes inclusivity by analyzing diverse datasets and integrating community input, ensuring that policies resonate with the varied needs of the population. It enables responsive resource allocation through real-time analysis of data patterns, ensuring efficient distribution that addresses immediate needs and adjusts to evolving priorities. In the proactive realm of crisis prevention, AI identifies potential scenarios through predictive analytics, empowering authorities to implement preventive measures. In addressing the intricate challenges of governance, AI excels by processing multifaceted data, providing nuanced insights for effective decision-making within adaptive governance. Furthermore, in times of crisis, AI facilitates adaptive responses by continuously analyzing real-time data, allowing authorities to tailor their strategies dynamically. Figure 4.1 indicates the smart systems for water resource management.

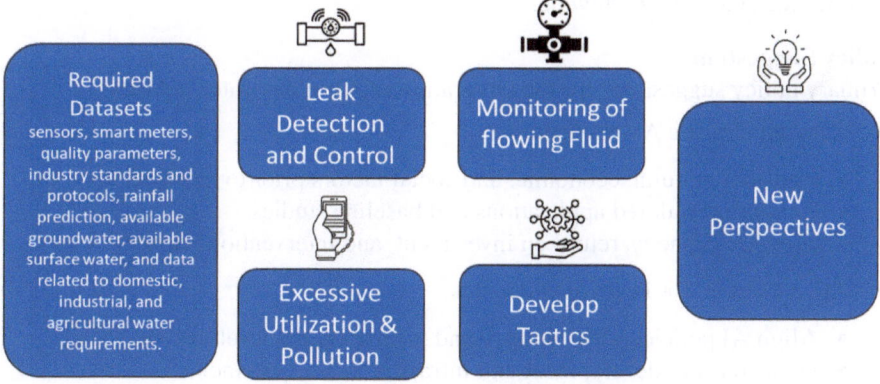

Fig. 4.1 Employing smart systems for water resource management or policy

4.5   Real-World Applications: AI Transforming Water Policy

Amidst the age of technological progress, the integration of artificial intelligence into water policy signifies a monumental shift. This article explores concrete, real-world instances where AI is reshaping water policy, fundamentally altering how we oversee, distribute, and safeguard our invaluable water resources. AI stands poised to propel the next wave of technological and economic advancement, akin to historical milestones such as the industrial revolution, the silicon chip era, and the advent of smart devices. Strategic foresight, which involves envisioning the future trajectory of an industry, is essential in informing present decision-making processes. This approach serves as a valuable tool in policy development, particularly in assessing AI's potential impact on achieving Sustainable Development Goals (SDGs) related to water. Current applications of AI in the water sector include: i) Predicting maintenance requirements for water infrastructure, ii) Forecasting water demand and consumption patterns, iii) Monitoring the ecological and environmental impacts of water reservoirs and dams, iv) Assessing water quality, and v) Detecting and predicting water-related disasters. These applications play a crucial role in advancing the objectives outlined in the Sustainable Development Agenda, particularly those related to health, access to clean water and sanitation, the development of sustainable cities and communities, and the preservation of terrestrial ecosystems. Table 4.1 represents the real-world AI-powered solutions worldwide.

Policy Suggestions
Primary policy suggestions for the integration of AI in the water industry.

1.   Conduct Holistic Assessments

   - Consider cultural, economic, and social factors prior to implementing AI.
   - Emphasize tailored applications and baseline studies.
   - Measure capacity, return on investment, and intervention impact.

2.   Align Policies for Positive Outcomes

   - Align AI policies with capacity and infrastructure development.
   - Focus on skill development and infrastructure requirements.
   - Sequence policies strategically for optimal impact.

3.   Invest in Workforce Development

   - Direct investments to educate a skilled workforce at all levels.
   - Offset job displacement and reduce reliance on the private sector.

4.   Collaborative Policies for Shared Challenges

   - Encourage collaborative policies at local and global levels.
   - Address cross-cutting water challenges for connected countries.

Table 4.1 Real-world applications across the world for AI in the water domain

AI-powered Solutions	Sources	Remarks
Detecting Pipe Bursts in Water Distribution Networks in Real Time	Romano et al. (2014)	UK
AI and Computer Vision for Automated Asset Condition Assessment	Myrans et al. (2018)	Used by companies in UK, Finland and Australia
Preventive Control of Wastewater Treatment Plants	https://aquasuite.ai/en/	The Aquasuite team is an integral part of Royal HaskoningDHV, a worldwide engineering, technology, and software company dedicated to advancing initiatives in the field of global clean water and sanitation
Intelligent Alerts for Proactive Management of Wastewater Networks	https://aquasuite.ai/en/products/flow/	Aquasuite FLOW represents a machine learning and AI innovation designed for wastewater transport. It seamlessly integrates precise forecasting of levels and flows with the automated and optimized control of pumping stations
Proactive Asset Management using Bayesian Networks	Economou et al. (2014), Babovic et al. (2002)	Experimental projects using this approach have been initiated in Stockholm, Singapore, the UK, and Denmark
Rainfall Monitoring using Computer Vision	Jiang et al. (2019), Allamano et al. (2015)	Organizations like WaterView in Italy, the Hydroinformatics Institute in Singapore, and academic institutions such as the Southern University of Science and Technology in Shenzhen, China, have adopted pragmatic approaches to address weather-related risks in fields including energy, automotive, and smart cities
Predicting short-term demand	Nassser et al. (2020)	Cairo
Identification and pinpointing of leaks	Shukla and Piratla (2020), Cody et al. (2020), Li et al. (2019), Zhou et al. (2019)	–
Identification of pollution	Li et al. (2019)	–
Real-time management like pump scheduling	Hajgat´o et al. (2020)	–

(continued)

Table 4.1 (continued)

AI-powered Solutions	Sources	Remarks
Identification and evaluation of issues in sewer systems	Hassan et al., (2019), Kumar et al. (2020), Wang et al. (2021)	–
Anticipation of water condition and forecasting of floods	Guo et al. (2021), Li et al. (2021)	–
Analyzing data from urban catchments, weather patterns, and flood data	Yang and Cervone (2019), Iqbal et al. (2021)	–
Managing floods and operating wastewater treatment facilities	Bowes et al., (2021), Hern´andez-del-Olmo et al. (2018)	–

5. Establish National Council or Agency

   - Establish a national-level representative council or agency.
   - Assist water agencies in adopting AI successfully.
   - Oversee the formulation of policies, guidelines, and codes of conduct.

6. Guiding Strategies for SDGs

   - Provide primary guidelines for crafting strategies to use AI.
   - Contribute to the fulfillment of SDGs related to water

Funding   This study was funded by the Department of Science and Technology, Government of India [DST/PRC/CPR/IITIndore (G)] for the project entitled "Technological Innovation and Intellectual Property"

References

Allamano, P., Croci A., and Laio, F. (2015), "Toward the camera rain gauge", *Water Resources Research*, Volume 51, Issue 3

Babovic, V., Drecourt, J.P.,Keijzer, M. and Hansen, P.F., (2002), "A data mining approach to modelling of water supply assets", *Urban Water*, 4 (4), 401–414

Bowes, B.D., Tavakoli, A., Wang, C., Heydarian, A., Behl, M., Beling, P.A., Goodall, J.L., 2021. Flood mitigation in coastal urban catchments using real-time stormwater infrastructure control and reinforcement learning. J. Hydroinformatics 23, 529–547. https://doi.org/10.2166/HYDRO.2020.080.

Cody, R.A., Tolson, B.A., Orchard, J., 2020. Detecting leaks in water distribution pipes using a deep Autoencoder and Hydroacoustic spectrograms. J. Comput. Civ. Eng. 34, 04020001 https://doi.org/10.1061/(asce)cp.1943-5487.0000881.

Economou, T., Bailey, T. and Kapelan, Z., (2014), "MCMC implementation for Bayesian hidden semi-Markov models with illustrative applications", *Statistics and Computing*, 24, 739–752.

Guo, Z., Leit˜ao, J.P., Sim˜oes, N.E., Moosavi, V., 2021b. Data-driven flood emulation: speeding up urban flood predictions by deep convolutional neural networks. J. Flood Risk Manag. 14, 1–14. https://doi.org/10.1111/jfr3.12684.

Hajgat´o, G., Pa´al, G., Gyires-T´oth, B., 2020. Deep reinforcement learning for real-time optimization of pumps in water distribution systems. J. Water Resour. Plan. Manag. 146, 04020079
https://doi.org/10.1061/(asce)wr.1943-5452.0001287.

Hassan, S.I., Dang, L.M., Mehmood, I., Im, S., Choi, C., Kang, J., Park, Y.S., Moon, H., 2019.
Underground sewer pipe condition assessment based on convolutional neural networks. Autom.
Constr. 106, 102849.

Hern´andez-del-Olmo, F., Gaudioso, E., Dormido, R., Duro, N., 2018. Tackling the start-up of a
reinforcement learning agent for the control of wastewater treatment plants. Knowl.-Based Syst
144, 9–15. https://doi.org/10.1016/j.knosys.2017.12.019.

Iqbal, U., Perez, P., Li, W., Barthelemy, J., 2021. How computer vision can facilitate flood management: a systematic review. Int. J. Disaster Risk Reduct. 53, 102030 https://doi.org/10.1016/j.
ijdrr.2020.102030.

Jiang, S., Babovic, V., Zheng, J, and Xiong, J., (2019), "Advancing opportunistic sensing in
hydrology: a novel approach to measuring rainfall with ordinary surveillance cameras",

Kumar, S.S., Wang, M., Abraham, D.M., Jahanshahi, M.R., Iseley, T., Cheng, J.C.P., 2020b. Deep
learning–based automated detection of sewer Defects in CCTV videos. J. Comput. Civil Eng.
34 (1), 04019047.

Li, D., Cong, A., Guo, S., 2019b. Sewer damage detection from imbalanced CCTV inspection data
using deep convolutional neural networks with hierarchical classification. Autom. Constr. 101,
199–208.

Li, Z., Liu, H., Luo, C., Fu, G., 2021. Assessing surface water flood risks in urban areas using
machine learning. Water (Switzerland) 13, 1–14. https://doi.org/10.3390/w13243520.

Myrans, J., Kapelan, Z. and Everson, R., (2018), "Automated detection of faults in sewers using
CCTV image sequences", *Automation in Construction*, vol. 95, 64–71.

Nasser, A.A., Rashad, M.Z., Hussein, S.E., 2020. A two-layer water demand prediction system
in urban areas based on micro-services and LSTM neural networks. IEEE Access 8, 147647–
147661. https://doi.org/10.1109/ACCESS.2020.3015655.

Romano, M. and Kapelan, Z., (2014), "Adaptive Water Demand Forecasting for Near Real-Time
Management of Smart Water Distribution Systems", *Environmental Modelling and Software*,
vol. 60, 265–276.

Shukla, H., Piratla, K., 2020. Leakage detection in water pipelines using supervised classification of
acceleration signals. Autom. Constr. 117, 103256 https://doi.org/10.1016/j.autcon.2020.103256.

Wang, M., Kumar, S.S., Cheng, J.C.P., 2021. Automated sewer pipe defect tracking in CCTV videos
based on defect detection and metric learning. Autom. Constr. 121, 103438.

Water Resources Research, 55 (4), 3004–3027

Yang, L., Cervone, G., 2019. Analysis of remote sensing imagery for disaster assessment using
deep learning: a case study of flooding event. Soft Comput 23, 13393–13408.

Zhou, M., Pan, Z., Liu, Y., Zhang, Q., Cai, Y., Pan, H., 2019a. Leak Detection and Location based
on ISLMD and CNN in a pipeline. IEEE Access 7, 30457–30464. https://doi.org/10.1109/ACC
ESS.2019.2902711.

Chapter 5
AI Framework for Future Water

Abstract The chapter presents a holistic framework aimed at incorporating AI into the evolution of water management practices. Acknowledging the growing complexities linked to global water resources, this framework delineates a strategic methodology to leverage the potential of AI. The abstract offers an outline of the primary elements constituting this framework, highlighting its ability to improve water efficiency, streamline resource allocation, and tackle emerging water-related issues. By examining the potential uses and impacts of this AI-centric framework, the chapter enriches discussions on constructing robust and eco-friendly water systems for the future.

5.1 Introduction to the AI Framework

As the global water crisis intensifies, with issues such as scarcity, pollution, and inadequate infrastructure, the need for innovative solutions becomes increasingly urgent. AI presents a glimmer of hope in this context, with the potential to address water-related challenges by analyzing large datasets, predicting trends, and refining decision-making processes. By embracing AI in water management, there is an opportunity to optimize resource allocation, enhance monitoring systems, and encourage conservation efforts. However, to fully harness AI's potential in the water sector, a comprehensive framework is crucial. This framework must guide the application of AI, ensuring its effectiveness in tackling current and future water challenges.

In the ever-changing landscape of water management, the integration of AI holds the potential to be a transformative force, providing innovative solutions to address the intricate challenges associated with water conservation and resource optimization. This section sets the stage for a comprehensive exploration of the future AI framework for water, exploring the potential applications, advantages, and shift in approach expected in how we manage one of our most valuable resources sustainably. The intersection of AI and water management offers the promise of efficiency,

M. K. Goyal et al., *AI Innovation for Water Policy and Sustainability*,
SpringerBriefs in Water Science and Technology,
https://doi.org/10.1007/978-3-031-72014-7_5

resilience, and intelligent decision-making, paving the way for a more sustainable and technology-driven water future.

5.2 Key Components of the AI Future Water Framework

In the expansive field of AI, an ongoing discourse revolves around the creation of "responsible" AI, ensuring its alignment with essential ethical values. To determine the extent of this discourse within the particular segment of AI literature centred on water, conducting a thorough assessment becomes essential. According to the article published by Doorn (2021), it is noted that very few papers have discussed cybersecurity in the realm of artificial intelligence (AI). This absence implies minimal discussion on the ethical dimensions of AI within the water domain. Over the last five years, non-academic institutions have increasingly formulated principles and guidelines for AI. The convergence on essential ethical principles is evident, yet there exists significant divergence in interpreting and implementing these principles across various domains, as highlighted by Jobin et al. (2019). In the literature, there is a focus on five fundamental principles when employing AI applications: transparency, justice and fairness, responsibility and accountability, privacy, and non-maleficence.

Figure 5.1 illustrates the five key principles—transparency, justice and fairness, responsibility and accountability, privacy, and non-maleficence governing the use of AI applications.

Table 5.1 presents the suggested drivers as per the report published by UNU-INWEH 2023 (Mehmood et al., 2020), the following are suggested for the adoption of AI for the sustainable development.

5.3 Applications of AI in Sustainable Water Practices

AI applications in sustainable water practices provide innovative tools to tackle challenges associated with resource conservation, quality preservation, and efficient management. This exploration delves into the diverse applications of AI, illuminating its transformative impact on sustainable water practices. From predictive analytics to advanced monitoring systems, AI offers a spectrum of solutions aimed at fostering a more resilient and environmentally conscious approach to water resource management. Across various sectors such as infrastructure, finance, marketing, and logistics, AI is transforming the operational landscape and reshaping interactions between businesses and customers. The emergence of potent and accessible tools such as ChatGPT, DALL-E, and Midjourney has demonstrated new frontiers for AI. Beyond traditional pattern recognition, AI now has the capacity to transcend boundaries and enhance its "creativity" by generating text or images based on simple text input, resembling human-like conversation. However, it is essential to emphasize the importance of fact-checking in this process.

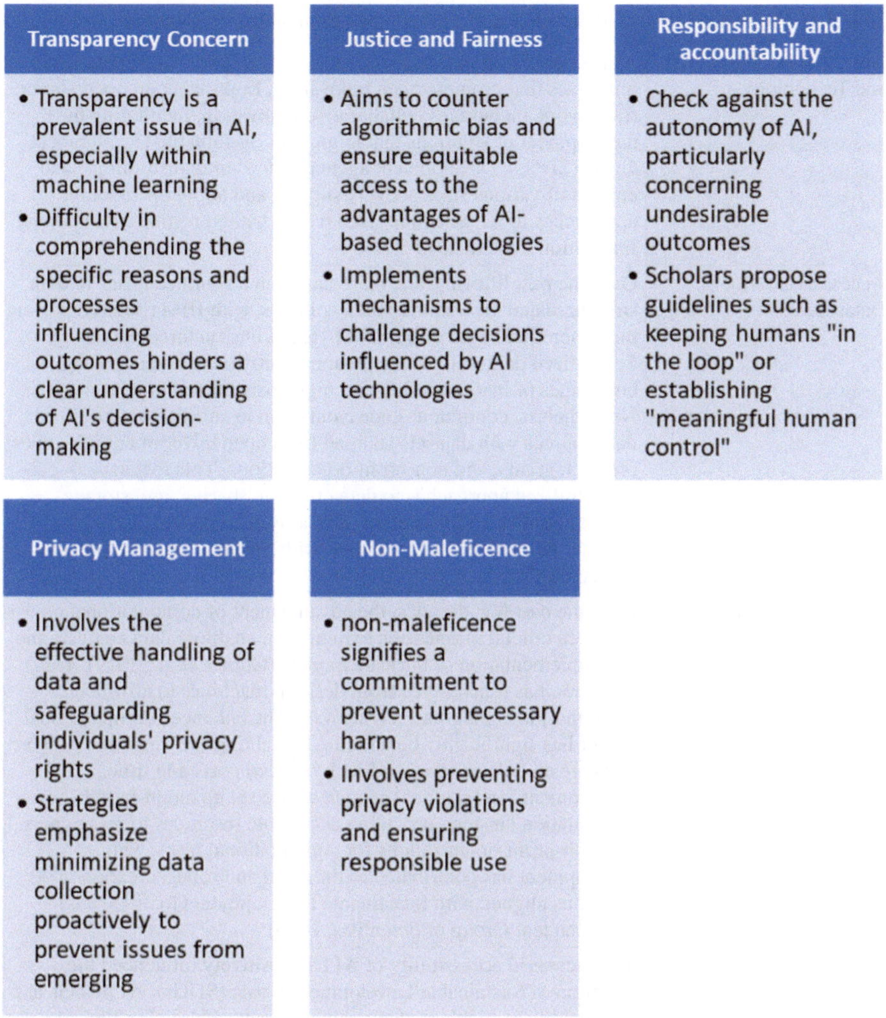

Transparency Concern	Justice and Fairness	Responsibility and accountability
• Transparency is a prevalent issue in AI, especially within machine learning • Difficulty in comprehending the specific reasons and processes influencing outcomes hinders a clear understanding of AI's decision-making	• Aims to counter algorithmic bias and ensure equitable access to the advantages of AI-based technologies • Implements mechanisms to challenge decisions influenced by AI technologies	• Check against the autonomy of AI, particularly concerning undesirable outcomes • Scholars propose guidelines such as keeping humans "in the loop" or establishing "meaningful human control"

Privacy Management	Non-Maleficence
• Involves the effective handling of data and safeguarding individuals' privacy rights • Strategies emphasize minimizing data collection proactively to prevent issues from emerging	• non-maleficence signifies a commitment to prevent unnecessary harm • Involves preventing privacy violations and ensuring responsible use

Fig. 5.1 Five key principles transparency, justice and fairness, responsibility and accountability, privacy, and non-maleficence governing the use of AI applications

In addressing challenges such as the escalating threat of extreme weather events, including flooding and rising sea levels, risks to our infrastructure become apparent. Numerous communities are vulnerable to the adverse effects of climate change. AI and emerging technologies play a pivotal role in tackling these challenges. Beyond conventional pattern recognition and historical data analysis, AI makes significant contributions that will reshape the landscape of water infrastructure planning and management in the future. Following are the five ways AI will revolutionize the future of water and infrastructure in the United States and globally.

Table 5.1 Suggested drivers for the adoption of AI for the sustainable development

Access to Infrastructure and Technology	AI frameworks mainly depend on open standards, open data, and initiatives that promote open innovation. Embracing an open-source framework encourages collaborative innovation, facilitating the development of solutions that align with Sustainable Development Goals (SDGs). Development agencies, academic institutions, and entities of various sizes actively support and adopt open-source approaches in AI, as emphasized by the United Nations Technology Innovation Labs in 2019
Access to Quality Datasets	Over the past fifteen years, there has been a significant rise in both structured and unstructured data volumes, with IBM (2013) reporting the generation of 2.5 quintillion bytes of unstructured data daily. Specialized data sources are predominantly centralized, posing challenges of inaccessibility and high costs for training AI models. Nevertheless, continuous endeavours aim to substitute centralized data sources with datasets sourced from open government initiatives, research grants, and non-profit organizations. This fosters decentralized approaches to data creation, sharing, and storage. These initiatives offer vital, high-quality seed data necessary for training AI models that are in line with Sustainable Development Goals (SDGs)
Access to Computing Power	Over the past few decades, the advancement of computational power has been crucial in scientific exploration, enabling data analysis and the implementation of intricate models (Gomes et al., 2019). This evolution has transitioned from desktop machines to affordable, plug-and-play cloud-based resources. The enhanced computational power has significantly benefited AI, enabling the execution of more intricate models on the cloud with reduced costs and time. Governments and regional agencies have set up cloud-based computation clusters, providing accessible resources to researchers and non-profit organizations for computational tasks. This development has contributed to the rapid and robust creation of AI solutions aligned with Sustainable Development Goals (SDGs) (Independent Group of Scientists, 2019)
Access to Complete AI Solutions	The increased accessibility of AI has positively influenced the adoption of Sustainable Development Goals (SDGs). AI models are now readily available in deployable formats, addressing typical challenges associated with AI technology adoption, such as the requirement for AI domain specialists, lengthy system integration processes, and transparency concerns (Dasgupta and Wendler, 2019)

5.3.1   Preventive Maintenance for Water Distribution

AI assists in predicting equipment malfunctions and required maintenance, improving operational uptime and reducing downtime. Several local municipalities and agencies in the United States have embraced AI solutions that can detect potential equipment failures in real-time, facilitating proactive maintenance measures to prevent issues.

5.3.2 Predicting Future Flood Risks

AI is transforming the assessment of future flood risk by utilizing advanced predictive modeling in ongoing nationwide pilot studies. By allowing AI to "learn" from both historical data and prior modeling, it enhances the ability to predict flood risks in complex areas with multiple risk factors. This proactive application of AI empowers authorities to take preemptive measures, safeguarding both infrastructure and citizens and minimizing the potential impact of flooding.

5.3.3 Quality of Water

AI's ability to analyze upcoming water quality trends, drawing insights from sensor data, helps identify shifts that may indicate contamination or related issues. This proactive approach allows local agencies to prepare preplanned action strategies, ensuring prompt responses to concerns like harmful algae blooms or other contaminants.

5.3.4 Sustainable Measures and Enhanced Energy Efficiency

AI improves energy efficiency in water treatment and distribution, resulting in cost savings and lower carbon emissions. Due to significant energy demands in these processes, AI provides optimization possibilities. Analysis of historical water usage patterns enhances anticipation of future demand, aiding in planning treatment and distribution networks. AI's optimization of the distribution network identifies areas of over- or underutilization by analyzing flow rates and pressure. This decision support system enables adjustments, enhancing water distribution efficiency and reducing energy consumption and carbon emissions.

5.3.5 Control and Utilization of Assets and Water Resources

AI contributes to the efficient handling and prioritization of infrastructure assets, guaranteeing their appropriate maintenance and replacement. Furthermore, it plays a role in water conservation by optimizing irrigation processes and minimizing unnecessary usage.

5.4 Challenges and Policy Recommendations in Implementing the AI Future Water Framework

The potential of Artificial Intelligence (AI) in water management is substantial, yet it faces several challenges. These include concerns about data privacy, the need for high-quality data, and financial implications associated with AI integration. Ensuring data privacy and security is crucial, especially given the extensive reliance on datasets, including sensitive information like personal water usage data. Effectively managing this data while respecting privacy rights poses a significant challenge in applying AI to water management. The effectiveness of AI in water management is closely tied to the quality of the data it utilizes. Flawed predictions or insights may result from inaccurate or incomplete data, emphasizing the challenge of maintaining consistent collection of high-quality, accurate data. The financial aspect of implementing AI technologies presents another significant challenge, particularly for smaller or less well-resourced water management organizations. Despite these challenges, the future of AI in water management is promising. Ongoing advancements in AI technologies are expected to enhance their capabilities, with more sophisticated algorithms improving prediction accuracy. Advances in data security technologies may address privacy concerns, and as the cost of AI technologies decreases, their adoption in water management is likely to become more feasible. The growing awareness of AI's potential benefits suggests continued growth in its utilization in water management. Table 5.2 provides policy suggestions for integrating Artificial Intelligence within the framework of water-related Sustainable Development Goals (SDGs).

5.5 Case Studies: Real-World Implementations of AI in Water Management

In this section, we explore the transformative impact of AI on water management, covering its practical applications and tangible outcomes across diverse contexts. Table 5.3 illustrates real-world case studies of AI implementations in water management.

AI has indeed made significant developments in various industries, including water management. The following section explores real-world examples of AI implementation in water management.

5.5.1 Chlorination and Water Quality Monitoring (Lowe et al., 2022)

AI is being used to optimize chlorination processes in water treatment plants. By analyzing historical data and real-time sensor readings, AI algorithms can adjust

Table 5.2 Policy suggestions for integrating Artificial Intelligence within the framework of water-related Sustainable Development Goals (SDGs)

Sl. No	Field	Policy Recommendations
1	Predictive Maintenance of Water Infrastructure	Local water agencies are required to follow regulatory frameworks set by provincial and federal governments, necessitating a thorough evaluation of costs and benefits related to leak detection. The gathered data should guide investment decisions. Moreover, provincial and federal governments should revise water management frameworks to incorporate advancements in water leak detection and prediction technologies offered by technology and engineering firms. This would contribute to both economic prosperity and environmental protection. It is recommended that governments enhance organizational capacity to enable local water agencies to deploy AI-based solutions for maintaining water infrastructure. Local governments are also encouraged to embrace data-driven approaches alongside AI to develop real-time sewage control and management tools and strategies
2	Predicting Water Demand and Consumption	The federal and provincial authorities ought to release and consistently refresh detailed instructions, along with benchmarked information and algorithms, for crafting AI-powered water management platforms. These platforms should possess the ability to forecast water demand and usage over both immediate and prolonged periods, spanning residential, industrial, and commercial domains. Water management entities are advised to amend their strategies to incorporate AI in predicting short and long-term water requirements and usage. Such forecasts are crucial for sustaining equilibrium between water demand and existing supplies, as well as for formulating efficient emergency strategies during severe weather occurrences. Urban water management blueprints should undergo evaluation and renewal every five years
3	Monitoring Water Reservoirs and Dams	Reservoir design regulations and directives should mandate the incorporation of an ample array of sensors and measuring instruments to fulfil present and future data requirements. This guarantees that AI algorithms can precisely forecast vital parameters indispensable for reservoir administration • To address transparency concerns associated with reservoirs, national and regional water management bodies should enforce policies promoting the adoption of open data and AI-driven instruments. This strategy facilitates the scrutiny of reservoir impacts pre and post inauguration • National and regional water management entities must bolster their organizational capacities to develop and deploy AI models for monitoring water reservoirs and dams

(continued)

Table 5.2 (continued)

Sl. No	Field	Policy Recommendations
4	Monitoring Water Quality	National water and health policymakers should establish and periodically update frameworks for monitoring water quality, integrating AI to predict water quality across different spatial scales (from large water bodies to households) and temporal scales (hourly, daily, weekly, and monthly). Policies should endorse the creation of open-access water quality databases at both national and regional levels, containing data on physical characteristics and colour patterns of all identified water contaminants. Furthermore, an incentive-based program should be introduced to incentivize local stakeholders to install and maintain measurement instruments
5	Monitoring and Forecasting Water-related Disasters	Disaster management agencies ought to formulate protocols for gathering and disseminating historical data concerning water-related calamities to bolster the precision of AI models. National authorities are encouraged to employ AI in generating regular risk maps derived from dynamic databases that reflect urban advancements

chlorine dosages more accurately, ensuring safe water quality. Additionally, AI models can predict water quality parameters (such as turbidity, pH, and dissolved oxygen) based on environmental factors, helping water utilities maintain optimal water quality.

5.5.2 River-Level Monitoring and Flood Prediction

AI-powered river-level monitoring systems use data from sensors placed along rivers to predict floods and issue early warnings. These systems analyze rainfall patterns, river flow rates, and historical flood data to provide timely alerts to communities and authorities.

For example, the Indian Institute of Technology (IIT) Roorkee developed an AI-based flood prediction model for the Ganga River, which has helped reduce flood-related damages.

Table 5.3 Case studies of AI implementations in water management

Implementations of AI	Sources
7Analytics provides real-time flood predictions for businesses and local authorities in Bergen, Norway. Utilizing expertise in hydrology, geology, and data science, their team of scientists develops highly accurate risk assessment tools to identify areas susceptible to flood damage. By considering not only weather patterns but also regional geography and drainage capacity, 7Analytics enables Norwegians to proactively plan for the current and future effects of climate change	https://7analytics.no/
To address the challenges encountered in restoring electricity supply in flooded areas, Neara, a London-based company, employs AI technology to automate LiDAR data analysis. This enables them to generate insights and create digital flood simulations specifically tailored for the electricity infrastructure sector. These digital models offer hyper-realistic details of various elements in the affected areas, including individual cable widths and the surrounding buildings and vegetation. Power networks can utilize these detailed simulations to make informed decisions aimed at minimizing damage and safely restoring power after flooding incidents. These decisions may include selectively deactivating cables, rerouting power, and identifying specific locations for engineers to intervene	https://neara.com/
The Flood Hub initiative by Google offers no-cost river flood notifications spanning over 80 countries, with a primary emphasis on India	https://sites.research.google/floods/l/0/0/3
A government agency in Australia, tasked with managing water supply and wastewater services in Southeast Queensland, initiated a project to enhance operational efficiency by modernizing its infrastructure. The main goals were to improve decision-making, enhance customer satisfaction, reduce costs, and promote sustainability. To achieve these objectives, the agency partnered with HCLTech to deploy a cloud-based, AI-driven intelligent data platform (IDP). This platform was designed to provide real-time insights and support data-driven decision-making processes	https://www.hcltech.com/case-study/revolutionizing-water-management-with-ai-driven-analytics
GramworkX has developed a smart farm resource management tool that utilizes both IoT and AI technologies to assist farmers in guiding, optimizing, and monitoring water usage. This device collects crucial farm parameters such as atmospheric temperature, pressure, humidity, rainfall, and soil moisture every 10 min and transmits this data to the cloud on an hourly basis. Subsequently, a machine learning algorithm generates irrigation and other predictions, which are accessible to farmers via a mobile application. To investigate irrigation requirements, the company conducted a comparative study across two regions cultivating the same crop (tomatoes). They also analyzed the growth patterns in two fields, one in Andhra Pradesh (AP) and the other in Maharashtra (MH), both utilizing drip irrigation and having similar acreage. The findings revealed that water requirements were higher in the MH region, a conclusion drawn from the modeling of temperatures and wind speeds	https://www.gramworkx.com/#/

5.5.3   Smart Irrigation and Aquaponics/Hydroponics Automation (Lowe et al., 2022)

AI-driven irrigation systems optimize water usage in agriculture. These systems consider soil moisture levels, weather forecasts, and crop requirements to deliver precise amounts of water.

In aquaponics and hydroponics, AI controls nutrient levels, pH, and lighting conditions for optimal plant growth. These techniques conserve water while maximizing crop yield.

5.5.4   Water-Quality-Index Modelling

AI models can calculate Water Quality Index (WQI) by aggregating various water quality parameters. WQI helps assess overall water quality and identify areas that need improvement.

Researchers have used AI to develop accurate WQI models for different water bodies, aiding policymakers in making informed decisions.

5.5.5   Desalination and Membrane Filtration

AI assists in optimizing desalination processes, making them more energy efficient. By analyzing data from desalination plants, AI algorithms improve membrane performance and reduce energy consumption. Membrane filtration systems, such as reverse osmosis, benefit from AI-driven predictive maintenance, ensuring uninterrupted water supply.

5.5.6   Community-Driven Water Management

Several case studies involve community participation in water management. Projects like rainwater harvesting, rural drinking water supply, and integrated watershed management engage local communities from the outset. Community involvement ensures sustainable practices and long-term success.

Furthermore, technological solutions can access dwindling water resources worldwide. Innovative approaches stemming from cost-effective innovations, yielding greater results with fewer resources, coupled with strategies focused on mimicking nature or biomimicry, hold promise for enhancing the limited freshwater resources of our planet (Raconteur, 2016). AI may also play a role in future advancements in

Table 5.4 Creative solutions to mitigate future water scarcity

Approaches	Uses
Water-efficient washing machine	Sheffield-based company Xeros has developed a device that reduces water consumption in washing machines by 80%, using polymer beads instead of traditional liquid detergents. Each machine has the potential to save over a million liters of water throughout its lifetime, equivalent to a decade's water consumption for an average UK household. Commercial machines, which typically use 500 L of water per cycle, benefit greatly from this technology. Additionally, this patented system, recognized with awards, reduces energy usage by up to 50% and requires approximately 50% less detergent compared to conventional systems. Xeros' innovative solution stands out as a notable success story in British exports
Portable water purification	The Solvatten organization, based in Sweden, has developed a portable water treatment and heating system tailored for off-grid households in developing regions. This system offers safe drinking water as well as hot water. Utilizing sunlight to eliminate harmful microorganisms and heat water to 75 °C, each unit serves approximately 250,000 people in tropical areas. Consisting of two five-liter containers, a fabric filter, and a heating unit, these units have a lifespan of around ten years despite costing up to $100 each. Solvatten is engaged in 45 active projects worldwide, reaching from Papua New Guinea to Mali
Biomimicry	Aquaporin, a Danish company, developed a groundbreaking water purification technology inspired by biomimicry. Their innovation features a membrane embedded with aquaporins, proteins found in cells, facilitating efficient filtration of pollutants and delivery of clean water. Using forward osmosis, this membrane requires significantly less energy than traditional methods. With 62 patents globally, it excels in treating wastewater, achieving remarkable efficiency—an impressive 2,700 L of water per second can be handled by just one gram of aquaporins
Converting waves into freshwater	A new device, the SAROS desalination buoy, harnesses wave power to convert seawater into drinking water. Developed by two University of North Carolina graduates, it uses reverse osmosis to purify brine and currently produces 500 gallons per day. Scaling up, it aims for 5,000 gallons per buoy, functioning in waves as low as 2.5 feet. US demonstration projects are set for early 2017, followed by pilots in Puerto Rico and Nicaragua. The innovation could offer clean water solutions for remote islands
Individual purification straw	LifeStraw, a portable water filter resembling a plastic straw, effectively removes pathogens and parasites from undrinkable water using hollow-fiber micro-filtration technology. With a capacity to purify up to 1,000 L without electricity or additional attachments, it's marketed by Danish company Vestergaard to outdoor enthusiasts. For each purchase, one schoolchild in a developing country receives safe drinking water for a year. Deployed in over 64 countries, it tackles water scarcity challenges effectively

(continued)

Table 5.4 (continued)

Approaches	Uses
Fog Collection	In Chile's Atacama Desert, where rain is scarce, fog presents an alternative water source. Local communities have created large mesh fog collectors using polypropylene, efficiently capturing water. Juan de Dios Rivera and Jacques Dumais are developing cost-effective, large-scale systems to provide water for entire communities, akin to modern wind turbines
Extracting water from the atmosphere	The WaterSeer utilizes condensation and wind power to generate up to 11 gallons of clean drinking water daily from the atmosphere, without electricity. It draws warm air through a metal tube underground, where water condenses and collects in a reservoir. VICI-Labs is crowdfunding to expand the concept for providing water to underserved communities

these areas. Table 5.4 illustrates approaches through which rising water scarcity can be mitigated.

AI plays a pivotal role in the collaborative effort to enhance water management, engaging researchers, policymakers, and local communities. The utilization of AI offers the potential to develop water systems that are more efficient, resilient, and sustainable. However, to ensure the effective integration of AI in water systems while avoiding the perpetuation of progress traps, three sets of commendations are proposed for the water industry. The first set of initiatives focuses on addressing infrastructure gaps and enhancing digital literacy, emphasizing the importance of upskilling the workforce and fostering partnerships with academic institutions and NGOs. Emphasis is placed on the importance of AI explainability to establish trust and facilitate interactions with end-users lacking technical expertise. The second set of recommendations proposes mechanisms to ensure trustworthy AI, including institutional practices such as red team exercises, bug bounties, and collaborative knowledge-sharing platforms. Software mechanisms involve collaborating with experts to set standards, develop interpretability guidelines, and implement user-testing methodologies. Hardware mechanisms concentrate on ensuring the capacity and reliability of physical resources. The third set advocates for the development of sector-specific legislation, regulations, and policies tailored to address the complexities of AI in water systems, encompassing technical standards, transparency, safety, accountability, security, human agency, and diversity and inclusion. In summary, a comprehensive approach is recommended, integrating educational, governance, and technological measures to guide the responsible implementation of AI across water systems (Richards et al., 2023).

In conclusion, the integration of AI is marking a transformative phase in water management, introducing novel approaches to improve water quality, preserve resources, and optimize infrastructure efficiency. In the face of ongoing challenges related to water scarcity and climate change, the significance of AI in fostering sustainable water management is poised to grow. Through the utilization of AI capabilities, we aspire to shape a future where our invaluable water resources are managed with efficiency and sustainability at the forefront.

Funding This study was funded by the Department of Science and Technology, Government of India [DST/PRC/CPR/IITIndore (G)] for the project entitled "Technological Innovation and Intellectual Property"

References

Dasgupta, A., & Wendler, S. (2019). AI Adoption Strategies, (9), 1–13. Retrieved from https://www.ctga.ox.ac.uk/files/aiadoptionstrategiesmarch2019pd

Doorn, N. (2021). Artificial intelligence in the water domain: Opportunities for responsible use. *Science of the Total Environment, 755,* 142561.

Gomes, C., Dietterich, T., Barrett, C., Conrad, J., Dilkina, B., Ermon, S., … Zeeman, M. Lou. (2019). Computational sustainability. Communications of the ACM, *62*(9), 56–65. https://doi.org/10.1145/3339399

https://iwa-network.org/wpcontent/uploads/2020/08/IWA_2020_Artificial_Intelligence_SCREEN.pdf

https://news.ucr.edu/articles/2023/04/28/ai-programs-consume-large-volumes-scarce-water

https://www.forbes.com/sites/federicoguerrini/2023/04/14/ais-unsustainable-water-use-how-tech-giants-contribute-to-global-water-shortages/?sh=316584a49392

https://www.wsp.com/en-us/insights/2023-artificial-intelligence-shaping-future-of-water

Independent Group of Scientists. (2019). The Future is Now: Science for Achieving Sustainable Development. United Nations.

Jobin, A., Ienca, M., & Vayena, E. (2019). The global landscape of AI ethics guidelines. *Nature machine intelligence, 1*(9), 389-399.

Lowe, M., Qin, R., & Mao, X. (2022). A review on machine learning, artificial intelligence, and smart technology in water treatment and monitoring. Water, 14(9), 1384.

Mehmood, H., Mukkavilli, S. K., Weber, I., Koshio, A., Meechaiya, C., Piman, T., ... & Liao, D. (2020). Strategic Foresight to Applications of Artificial Intelligence to Achieve Water-related Sustainable Development Goals. *United Nations University Institute for Water, Environment and Health, Hamilton, Canada. UNU-INWEH Report Series.*

Raconteur. (2016). Future of Water. 424, 16. Retrieved from http://rcnt.eu/wn0

Richards, C. E., Tzachor, A., Avin, S., & Fenner, R. (2023). Rewards, risks and responsible deployment of artificial intelligence in water systems. *Nature Water*, *1*(5), 422–432.